Rural Transformations

This book focuses on the transformation of rural places, peoples, and land endemic to the contemporary manifestations of globalization.

Migration, global economic restructuring, and climate change are rapidly transforming rural places across the globe. Yet, global attention characteristically focuses on urban social and economic issues, neglecting the continued roles of rural people and places. Organized around the three core themes of demographic change, rural–urban partnerships and innovations, and landscape change, the case studies included in this volume represent both the Global North and Global South and underscore the complexity and multi-scalar nature of these contemporary challenges in rural development, planning, and sustainability.

This book would be valuable supplementary reading for both students and professionals in the fields of rural land management and rural planning.

Holly Barcus is a DeWitt Wallace Professor of Geography at Macalester College (USA). Her interests reside at the intersection of migration, ethnicity, and rural peripheries. For the past 15 years she has been working in western Mongolia amongst the Kazakh population considering questions of identity, environment, and changing migration trajectories. She holds positions on the editorial board for the *Journal of Rural Studies* and as a co-chair of the International Geographical Union's Commission on the Sustainability of Rural Systems (IGU-CSRS).

Roy Jones is an Emeritus Professor of Geography at Curtin University in Perth, Western Australia, where he has worked since moving to Australia in 1970. He is an historical geographer with particular interest in the areas of rural and regional change. In 2013, he was awarded a Distinguished Fellowship of the Institute of Australian Geographers.

Professor Serge Schmitz teaches rural geography, tourism strategy, regional development, and landscape planning at the University of Liège (Belgium) and leads the Laboratory for the Analysis of Places, Landscapes and European countryside (Laplec), since 2007. Early work focused on land consolidations, natural parks, and landscape analyses. Today, his research focuses on multifunctional countryside, in Wallonia and around the world, with a special interest for heritage landscapes, rural tourism, and ways of dwelling.

Perspectives on Rural Policy and Planning

This well-established series offers a forum for the discussion and debate of the often-conflicting needs of rural communities and how best they might be served. Offering a range of high-quality research monographs and edited volumes, the titles in the series explore topics directly related to planning strategy and the implementation of policy in the countryside. Global in scope, contributions include theoretical treatments as well as empirical studies from around the world and tackle issues such as rural development, agriculture, governance, age and gender.

Service Provision and Rural Sustainability
Infrastructure and Innovation
Edited by Greg Halseth, Sean Markey and Laura Ryser

The Changing World of Farming in Brexit UK
Edited by Matt Lobley, Michael Winter and Rebecca Wheeler

Rural Gerontology
Towards Critical Perspectives on Rural Ageing
Edited by Mark Skinner, Rachel Winterton and Kieran Walsh

Tourism and Socio-Economic Transformation of Rural Area
Evidence from Poland
Edited by Joanna Kosmaczewska and Walenty Poczta

Governance for Mediterranean Silvo-Pastoral Systems
Lessons from the Iberian Dehesas and Montados
Edited by Teresa Pinto-Correia, Helena Guimarães, Gerardo Moreno and Rufino Acosta Naranjo

Politics and Policies of Rural Authenticity
Edited by Pavel Pospěch, Eirik Magnus Fuglestad and Elisabete Figueiredo

Rural Transformations
Globalization and Its Implications for Rural People, Land, and Economies
Edited by Holly Barcus, Roy Jones and Serge Schmitz

For more information about this series, please visit: http://www.routledge.com/Perspectives-on-Rural-Policy-and-Planning/book-series/ASHSER1035

Rural Transformations

Globalization and Its Implications for Rural People, Land, and Economies

Edited by Holly Barcus, Roy Jones, and Serge Schmitz

Routledge
Taylor & Francis Group

LONDON AND NEW YORK

First published 2022
by Routledge
4 Park Square, Milton Park, Abingdon, Oxon OX14 4RN

and by Routledge
605 Third Avenue, New York, NY 10158

Routledge is an imprint of the Taylor & Francis Group, an informa business

British Library Cataloguing-in-Publication Data
A catalogue record for this book is available from the British Library

Library of Congress Cataloging-in-Publication Data
Names: Barcus, Holly R., editor. | Jones, Roy, 1946- editor. | Schmitz, Serge, editor.
Title: Rural transformations : globalization and its implications for rural people, land, and economies / edited by Holly Barcus, Roy Jones and Serge Schmitz.
Description: 1 Edition. | New York, NY : Routledge, 2022. | Series: Perspectives on rural policy and planning | Includes bibliographical references and index.
Identifiers: LCCN 2021045517 (print) | LCCN 2021045518 (ebook) | ISBN 9780367626464 (hardback) | ISBN 9780367626471 (paperback) | ISBN 9781003110095 (ebook)
Subjects: LCSH: Rural development. | Land use, Rural. | Sociology, Rural. | Globalization--Social aspects. | Globalization--Economic aspects.
Classification: LCC HN18.3 .R857 2022 (print) | LCC HN18.3 (ebook) | DDC 307.1/412--dc23
LC record available at https://lccn.loc.gov/2021045517
LC ebook record available at https://lccn.loc.gov/2021045518

ISBN: 978-0-367-62646-4 (hbk)
ISBN: 978-0-367-62647-1 (pbk)
ISBN: 978-1-003-11009-5 (ebk)

DOI: 10.4324/9781003110095

Typeset in Goudy
by SPi Technologies India Pvt Ltd (Straive)

Contents

List of figures	viii
List of tables	x
List of contributors	xi
Acknowledgements	xv

1 Introduction 1
HOLLY BARCUS, ROY JONES, AND SERGE SCHMITZ

PART I
Agricultural and Land Use Transitions 13

2 Agribusiness Towns, Globalization and Development in Rural Australia and Brazil 15
MICHAEL WOODS

3 A Change in the Role of Women in the Rural Area of Southeastern Anatolia, Turkey 32
SUK-KYEONG KANG

4 Agricultural Transition in Rural China: Intersections of the Global, National and Local 53
GUY M ROBINSON, BINGJIE SONG, AND ZHENSHAN XU

5 A Checkered Pathway to Prosperity: The Institutional Challenges of Smallholder Tobacco Production in Zimbabwe 71
TARIRO KAMUTI

PART II
Demographic Diversity

89

6 The Changing Rural Periphery: Contested Landscape,
Agricultural Preservation, and New Rural Residents in
Dakota County, Minnesota, U.S.A.   91
HOLLY BARCUS AND DAVID A. LANEGRAN

7 Labour Immigration and Demographic Transformation:
Lithuanian and Polish Nationals in Rural Ireland   113
MARY CAWLEY

8 Shaping Public Spaces in Rural Areas: Lessons from
Villages in the Gmina of Krobia, Poland   132
KAROLINA DMOCHOWSKA-DUDEK, MARCIN WÓJCIK,
PAULINA TOBIASZ-LIS, AND PAMELA JEZIORSKA-BIEL

9 Mother's Little Helper: A Feminist Political Ecology of
West Africa's Herbicide Revolution   151
WILLIAM G. MOSELEY AND ELIZA J. PESSEREAU

PART III
Rural Innovations and Urban–Rural Partnerships

167

10 Towards a Strategic Model for Sustainable Agriculture
in Mediterranean Countries: A Case Study of the
Cooperativa Hortec (Catalonia, Spain)   169
JOAN TORT-DONADA AND JORDI FUMADÓ-LLAMBRICH

11 Rural Innovation and the Valorization of Local Resources
in the High Atlas of Marrakesh   186
FATIMA GEBRATI

12 Does an Agricultural Products' Certification System
Reorganize Vegetable Farmers? A Case of the VietGAP program
in Lam Dong Province, Vietnam   200
DOO-CHUL KIM, TUYEN THI DUONG, QUANG NGUYEN,
AND HUNG THE NGUYEN

13 Relocalizing Food Systems for Everyone, Everywhere?
Reflections on Walloon Initiatives (Belgium)   217
ANTONIA BOUSBAINE AND SERGE SCHMITZ

14 Conclusion   231
HOLLY BARCUS, ROY JONES, AND SERGE SCHMITZ

Index   239

Figures

2.1	Change in farmland use in Dom Pedrito municipality, 2002–2018	20
2.2	The VLDC Woolnorth property near Smithton, acquired by Moon Lake Investments in 2016	24
2.3	Soybean fields encroaching on cattle pasture near Dom Pedrito	27
2.4	The dairy farming landscape near Smithton	28
3.1	NUTS* statistical regions of Turkey	33
3.2	Location of study area (Çukurkuyu town, Niğde Province, Turkey)	37
3.3	Respondents' seasonal migration routes	39
3.4	Temporary living conditions for families engaged in seasonal agricultural work	42
3.5	Living conditions of the families of seasonal agricultural workers in their hometowns	42
3.6	The educational levels of female workers ($N = 55$)	44
3.7	Number of children in respondent families ($N = 29$)	44
3.8	Actual daily wages of female workers	47
3.9	Economic independence as an indicator of the status of women	48
4.1	The study area: White Deer Plain, Shaanxi Province, North-west China	58
4.2	Grapes being grown by smallholders for JunDu organic grapes in White Deer Plain	61
4.3	Bailuyuan – Bailucang scenic area	64
5.1	Miombo woodlands and tobacco growing areas	72
5.2	A conventional tobacco barn and logs of firewood from indigenous trees	80
6.1	Percent change in non-White population, 2000–2017	97
6.2	Percent change in Latinx population, 2000–2017	97
6.3	Buddhist Templein Dakota County, Minnesota	99
6.4	Simpson Index of Diversity of farm operators, 2012	104
7.1	Ireland: counties	117
7.2	Ireland: distribution of Lithuanian nationals, 2016	122
7.3	Ireland: distribution of Polish nationals, 2016	123
7.4	Ireland: distribution of clusters of EDs	127
8.1	Locations of the villages studied within the gmina of Krobia	133

8.2	The spatial development pattern in Domachowo: A – in 1933, B – in 2018	136
8.3	Centres of the village of Domachowo	137
8.4	The spatial development pattern in Stara Krobia: A – in 1933, B – in 2018	139
8.5	The spatial development pattern in Potarzyca: A – in 1933, B – in 2018	140
8.6	Biskupian folk culture in the public space of Domachowo (2018)	140
8.7	Theory of change map for the shaping of action over public spaces in rural areas	144
9.1	Map of field sites in Southwestern Burkina Faso	155
9.2	Atrazine (left) and glyphosate (right) are commonly used herbicides in Burkina Faso	156
9.3	Herbicide seller at the village level (left) and male relative of female farmer applying herbicides in her rice field (right)	159
10.1	Image of a traditional Catalan farmhouse, or *masia*, in the eastern Pyrenees	173
10.2	Mixed agricultural landscape in El Maresme, in the north-east corner of the Barcelona metropolitan region	176
10.3	Location of the Catalan agricultural holdings directly linked to *Hortec* (2020)	178
10.4	*Hortec* logo	180
11.1	The Study Area	190
11.2	Economic Interest Grouping (EIG) in the L'Oued Zat Basin	194
12.1	Location of the study site	203
12.2	Changes in the VietGAP certified area in the Duc Trong district	205
12.3	Access to the VietGAP cultivation	209
13.1	Location map	218

Tables

3.1 Seasonal agricultural activities conducted by the respondents
 in the summer of 2017   40
3.2 Average daily wages of seasonal agricultural workers (TRY*)   47
4.1 The agricultural profile of sample villages in White Deer Plain   58
4.2 Characteristics of cherry production in four sample villages   61
5.1 Registered tobacco growers according to sector   77
5.2 Flue-cured tobacco production 2000–2018   78
6.1 Regional population growth, 1900–2017   96
6.2 New rural populations in Dakota County as represented through
 changing representations of religious entities   99
6.3 Percent change in farm operators, 2002–2017   103
7.1 Ireland: population usually resident and present on census night,
 by nationality   115
7.2 Irish, Lithuanian and Polish nationals: % distribution between
 different town size groupings and rural areas   116
7.3 Hoover Index values of population concentration of nationality
 groups for different Irish geographical areas, 2006, 2011 and 2016   124
7.4 Index of dissimilarity values for Irish nationals and Polish and
 Lithuanian nationals for different geographical areas, 2006,
 2011 and 2016   125
7.5 Results of cluster analysis of 'rural' electoral districts: average
 variable values   126
9.1 Herbicide application help by women's age   161
12.1 Farm characteristics   206
12.2 Contract and price guarantee   207
12.3 Crop choice decision and harvesting process   207
12.4 Technical advice for vegetable cultivation   208
12.5 Revenue change among the VietGAP households/farmers   208
12.6 Farmers' perceptions on the necessary conditions to become a
 VietGAP farmer   210

Contributors

Antonia Bousbaine is a high school teacher in Tournai (Belgium) and a scientific collaborator of Laplec at the University of Liège (Belgium). Her doctoral research investigates the impact of the growing demand for local food on the relationships between medium-sized cities and the surrounding countryside.

Mary Cawley is an Adjunct Professor in the Whitaker Institute for Innovation and Societal Change and Senior Lecturer Emerita in the School of Geography, Archaeology and Irish Studies, at the National University of Ireland, Galway. Her research and publications, individually and with colleagues, relate to the geography of rural societal change and sustainability, notably aspects of agriculture, population, service provision and tourism. Ongoing research includes study of the changing composition of rural populations through international labour immigration.

Karolina Dmochowska-Dudek is an Assistant Professor in the Department of Regional and Social Geography at the University of Lodz (Poland). Her research interests within social and rural geography include geomatics and locational conflicts, especially the NIMBY syndrome which arises from the differing needs of increasingly diversified local communities.

Tuyen Thi Duong participated in the Okayama – Hue international master's program on the sustainability of rural and environmental systems and received her master's degree in environmental science in 2018. After returning to Vietnam and Hue University where she now lectures in the Faculty of Economics and Development Studies. She currently conducts research on economic development in relation to the protection of the natural environment and sustainable livelihoods in rural areas.

Jordi Fumadó-Llambrich obtained a master's degree in environmental sciences from the University of Barcelona in 2014. He works in the private sector as a technical advisor on organic farming and is undertaking doctoral research on agricultural cooperatives in Catalonia.

Fatima Gebrati is a Professor-researcher in the Department of Geography at the University Cadi Ayyad, Marrakesh, Morocco. Her research interests include regional development and political geography. She teaches in a range of specialized master's degree programmes in Morocco, France, and South Africa. She researches in the areas of rural development, tourism, land use planning, development studies, borders, and migration.

Pamela Jeziorska-Biel is an Assistant Professor in the Department of Local Government Economics at the University of Lodz (Poland). Her main research interests encompass issues of local development, village renewal and social research methodology. Currently, she is a participant in the HORIZON 2020 project: "Resituating the Local in Cohesion and Territorial Development – RELOCAL". She is a member of the Polish Geographical Society and is active in the Commission of Rural Areas.

Tariro Kamuti is a Research Fellow in the Center for Gender and Africa Studies at the University of the Free State in Bloemfontein, South Africa. His research interests include governance issues encompassing sustainable utilization of natural resources, the social dimensions of biodiversity conservation and sustainable rural and urban development. He is a member of the International Society of Sustainability Professionals USA, the Regional Studies Association UK, and the Society of South African Geographers.

Suk-Kyeong Kang is an Assistant Professor in the Geography Department at Niğde Ömer Halisdemir University, Turkey. Her current research interests focus on the revitalization of rural areas, the livelihoods of seasonal migrant workers, and the transformation of rural areas resulting from the influx of Syrian refugees into Turkey.

Doo-Chul Kim is a Professor in the Graduate School of Environmental and Life Science, Okayama University, Japan. He researched at the United Nations Centre for Regional Development (UNCRD) from 1992 to 1994 before joining Tohoku University and, in 1999, Okayama University. His current research interests are natural resource management in upland Vietnam and comparative studies of the sustainability of rural systems and environmental controversies in rural areas in Vietnam and East Asia.

David A. Lanegran is the John S. Holl Professor of Geography, Emeritus at Macalester College. As a distinguished scholar of geography, including historic preservation and urban planning, he was President of the National Council for Geographic Education (NCGE) in 1998 and received the 2009 NCGE George J Miller Award and 2010 Gilbert Grosvenor Honors for Geographic Education from the American Association of Geographers.

William G. Moseley is DeWitt Wallace Professor of Geography and Director of the Program for Food, Agriculture and Society, at Macalester College. His research interests include political ecology, tropical agriculture, environment and development policy, livelihood security, and Africa.

Hung The Nguyen is a Lecturer at Hue University of Economics, Vietnam. He is also a doctoral candidate at the Graduate School of Environmental and Life Science, Okayama University, Japan. His current research is on the local impacts of the global food supply chain, focusing on Vietnamese catfish.

Quang Nguyen is a Lecturer in the School of Economics at the University of Economics Ho Chi Minh City, and a research fellow at Environment for Development (EfD)-Vietnam, an environmental economics research centre of the EfD network. His research interests include rural transition, agriculture, and environmental problems. His current work focuses on livelihood and land use change in peri-urban areas and the evaluation of wetland ecosystems in Vietnam.

Eliza J. Pessereau graduated from Macalester College in 2019 with BAs in biology and geography. While at Macalester she studied abroad in Madagascar, and later served as an agriculture extension volunteer with the Peace Corps in Cameroon. Her research interests include agroecology, food insecurity, foraging, plant development, and the climate crisis.

Guy M. Robinson is a Professor in the Department of Geography, Environment & Population, University of Adelaide, South Australia and a Departmental Associate in the Department of Land Economy, University of Cambridge, UK. He has worked on rural development and environmental management for 40 years, with previous appointments at the Universities of Oxford, Edinburgh, Kingston London and South Australia and visiting positions at various institutions in Australia, Canada, China, India, New Zealand, and the UK.

Bingjie Song is a PhD candidate in the Department of Geography, Environment and Population, at the University of Adelaide, South Australia. Her research interests mainly focus on multifunctional agriculture and climate change adaptation in Australia and China, and agricultural transformation in China post-1978.

Paulina Tobiasz-Lis is an Assistant Professor at the Department of Regional and Social Geography of the University of Lodz (Poland). Her main research interests have been focused on socio-spatial changes of urban and rural areas and their perceptions. She has participated in several projects of national and international scope.

Joan Tort-Donada has been a Professor of Geography at the University of Barcelona since 1990. He has worked on regional analysis, spatial planning, geopolitics and border studies, toponymy, landscape studies, cultural and historical geography, geographical thought, and literary geography.

Marcin Wójcik is a Professor at the University of Lodz (Poland), Faculty of Geographical Sciences, Department of Regional and Social Geography. He has authored publications on rural development, cultural landscape, local development, and socio-spatial diversities, and is a manager of national and international projects.

Michael Woods is a Professor of Human Geography at Aberystwyth University in Wales and Co-Director of the WISERD/Civil Society research centre. He studies changing rural communities and recently completed the European Research Council project GLOBAL-RURAL, investigating globalization in 15 rural localities. He has also researched rural–urban relations, rural protests and politics, globally engaged farmers, and rural civil society.

Zhenshan Xu is a PhD student in the Management School at Waikato University, New Zealand where he is working on the development of theme parks in China. He has a master's degree in Tourism Management from Shaanxi Normal University, Xi'an, China.

Acknowledgements

The development and completion of an edited volume such as this is the culmination of many ideas and sources of support. We, the editors, would like to sincerely thank all of our contributors who have worked closely with us to produce this volume. As we note in the Introduction, the research papers profiled in this volume were first presented at the 27th Annual Colloquium of the International Geographical Union Commission on the Sustainability of Rural Systems, co-hosted by the Geography Department at Macalester College and the Geography and Anthropology Department at the University of Wisconsin Eau Claire (UWEC), in Minnesota and Wisconsin, USA, respectively. Both Macalester and UWEC faculty, staff, and students played key roles in organizing the Colloquium and we are sincerely thankful for this support. We would also like to thank Provost Karine Moe, Macalester College, for providing funding and in-kind support for the Colloquium, including resources for facilities and other activities related to the Colloquium, including this edited volume. Without her financial support and enthusiasm for gathering 46 scholars from 16 countries around the world, much of this work would not have been possible. Further significant support was provided by the National Science Foundation (Award 1853832), and we are grateful for this important contribution to the success of the Colloquium and this edited volume.

1 Introduction

Holly Barcus, Roy Jones, and Serge Schmitz

Introduction

The contemporary global rural landscape is comprised of highly varied and multidimensional social, economic and physical geographies. The forces of change influencing rural places are local, national, regional and international in scale and inclusive of historic and contemporary political economic and social processes. As Woods notes, "....in identifying globalization as one of the pre-eminent forces of our time, globalization is conceptualized not as the movement of goods, people and capital around the world, but as the advanced interconnection and interdependence of localities across the world..." (2005, 32), such interconnections and interdependencies challenge the notion of a singular perspective on rural change and deepen our need to begin assessing how rural transformations are occurring across the global landscape.

This edited volume is concerned with understanding the multidimensional, globally diverse spatial manifestations of rural change. Each chapter was originally contributed to the July 2019 27th Annual Colloquium of the International Geographical Union Commission on the Sustainability of Rural Systems, held in Minnesota and Wisconsin, USA. The Colloquium theme was "Sustaining Rural Systems: Rural Vitality in an Era of Globalization and Economic Nationalism," with each of the sub-themes represented in this edited volume. From the many excellent presentations at the Colloquium, we sought, in this volume, to include a selection of chapters that were geographically and thematically representative and which, collectively, could capture some of the diversity of global rural transformations taking place. The authors whose research is profiled here embrace a diversity of approaches in examining contemporary rural change. Case studies were selected to profile this diversity. Thirteen countries are represented, with most authors coming from institutions within these respective countries. Some chapters offer greater theoretical insights to change, while others adopt more empirical perspectives. Despite ongoing, often nation state-based, debates on the definition of rurality, we intentionally do not adopt a specific definition of rural or rurality but rather leave these distinctions to the authors.

We also adopt an international perspective on rural change. While, in some spheres, discussion of rural changes occurring in "developed market economies" is segregated from discussion of rural change elsewhere across the globe, we argue

DOI: 10.4324/9781003110095-1

that rapid and widespread processes of globalization have made change in rural places dynamic and highly varied, and that the acceptance of a Global North and Global South dichotomy requires increased scrutiny. This dichotomy fails to account for the vast continuum of change in rural places and the need to acknowledge the similarities and differences in experiences inherent in and across rural places. We argue for a more inclusive and far-reaching acknowledgement of rural change and its dynamic connections across, between and within rural places in the Global North, Global South and Global Middle.

The remainder of this brief introductory chapter is divided into a consideration of our three core themes: (1) agricultural and land use transitions, (2) demographic diversity and (3) rural innovations and urban–rural partnerships. In each section, we first discuss overarching issues and trends related to each theme followed by specific highlights from each of the chapters that comprise the theme.

Part I: Agricultural and Land Use Transitions

Until relatively recently, most land use transitions resulted from the conversion of natural biomes to agricultural or pastoral land, a shift which began with the initial domestication of plant and animal species by humans approximately 10,000 years ago. This process, which first commenced in the river valleys of the Middle East and Asia, but also developed independently in parts of the Americas, first extended to Europe, where large areas of forest were cleared and replaced with farmland in the Middle Ages. The expansion of the earth's agricultural and pastoral areas reached its zenith in the age of imperialism from the sixteenth to the early twentieth centuries when European entrepreneurs and settlers facilitated agricultural expansion both within and beyond their overseas colonies (Belich 2009; Jones and Diniz 2019).

Throughout this period, the primary motive for land use change was production. Ensuring the supply of ever greater volumes of food and fibre products for a growing population required ongoing land clearance and agricultural expansion. In recent decades, however, the earth's rural areas have been experiencing a varied set of transformations that result from a much wider range of motives. This transition, which was initially identified as a shift from productivism to post-productivism (Wilson 2001; Roche and Argent 2015), has varied purposes and is generating complex changes in and to a countryside which is becoming much more diverse both functionally and spatially. Holmes (2006, 144) explains this shift in terms of a three-way "revaluation of rural resources" based not only upon production but also on consumption and protection.

In many of the world's rural areas, agricultural production remains the primary focus. However, mechanization and, potentially, automation are bringing about what Holmes terms "agricultural hyper productivity" whereby more and more products can be obtained from a given area of land with ever-decreasing labour inputs. The resultant job losses cause people either to leave rural areas for cities in search of work or to look for alternative or additional sources of employment and income in their rural locations. These remaining productivist rural areas currently face further uncertainties because many now supply their products to distant markets through global supply chains. They are therefore vulnerable to price shifts and international competition as well as to increasingly variable weather conditions in a time of climate change. Soil degradation is another important challenge. Even when these rural regions retain their productive capacity, agricultural transitions can occur as a result of technological, market or climatic changes.

The technological advances which have allowed urban dwellers to source agricultural produce from distant parts of the world now also allow people to consume the countryside as tourists, as long-distance commuters, as retirees and even as "electronic cottagers" (Gold 1991) who can work from home using digital technologies. This revaluation of the countryside as a potential site of consumption rather than production can bring inflows of people and income that can counter the effects of agricultural mechanization and rural depopulation. But this process of "cashing out" of the city and "cashing in" to the countryside (Curry et al. 2001) is spatially selective. The rural areas that are desirable and demanded for their consumption values need to possess specific amenity characteristics including relative proximity to larger towns, scenic terrain and cultural landscapes, temperate climates and, ideally, an existing population of creative and like-minded inhabitants (Argent et al. 2014).

Overlapping with and complementing this valorization of rural areas for consumption purposes is a growing social awareness of the need to protect aspects, or at least examples, of both natural and cultural environments within rural areas. The relatively recent increase in the importance of these protection values has both scientific and heritage components. The destruction of natural environments, and particularly of forests, has prompted concerns over biodiversity loss and even climate change and has led to the growth of global and local environmental conservation movements (Jacoby 2014). In parallel with this, local communities, nation states and even international bodies such as UNESCO have increasingly sought to protect both landscapes and cultures in the name of heritage (Lowenthal 1997). Again, only certain rural areas are seen as meriting high protection values as a result of, for example, the presence of threatened species of flora and fauna, spectacular scenery, idiosyncratic and historic agricultural landscapes or sites of significance to Indigenous groups.

Given the presence of what are now the global and globalizing "driving forces" of production, consumption and protection, Holmes contends that rural areas are transitioning from a predominantly productivist past to an increasingly multifunctional present. Furthermore, he argues that these three forces interact and, on occasion, compete resulting in "increasing diversity, complexity and spatial heterogeneity in modes of rural occupance" (2006, 144). Some rural areas will remain primarily dependent on productivist agriculture as a result of their suitable climates, soils, transport links and inputs of human and financial capital, but they will be continuously transformed by technological change and by the vicissitudes of global markets and politics. Rural areas dependent on consumption activities are equally dynamic. The COVID-19 pandemic may be constraining international tourism, a major employer in many parts of the rural world, but it is also accelerating the growth of work, including highly remote work, from home allowing more people to move from urban to rural areas. Areas designated for their protection values may also be subject to rapid change. The late twentieth and early twenty-first centuries saw the designation of many national and regional parks, World Heritage sites and Indigenous protected areas. But recent political decisions placing production values above protection values have reduced or removed the protected status of numerous environmental conservation areas and Indigenous lands in both Brazil and the USA.

The agricultural and land use transitions that are currently taking place across the rural world are the results of a diverse range of, often competing, technological, cultural, economic and environmental trends that operate at different levels of intensity and in different locations. They therefore need to be studied in local contexts in order to obtain the understandings offered by the case studies presented in this volume. To this end, the case study chapters in this section focus upon both the causes and the implications of agricultural and land use change, drawing upon examples from four continents.

Woods documents the changes in agriculture and land use taking place in the "agribusiness towns" (Pequeno and Elias 2015) of Dom Pedrito in Rio Grande do Sul, Brazil and Smithton in Tasmania, Australia and their hinterlands. He demonstrates how not only the surrounding rural areas but the two towns themselves have been transformed by transnational agribusiness corporations into "global farmlands" drawing in capital and technology, including seeds, pesticides and fertilisers, from these corporations and supplying new forms of produce to international markets. While this change process has augmented the productivity of local agriculture in both locations, it has also disrupted pre-existing economic and social linkages and brought about increased income inequality and environmental degradation.

Robinson, Song and Xu also interrogate a driving force for change in their study of the modernization of an agricultural region close to the North-Western Chinese city of Xi'an. Here, it is capitalism with Chinese characteristics, rather than global capital, that is bringing about radical agricultural and land use change. In recent decades, the process of urban and industrial development in Xi'an has led to the growth of consumer demands from an increasingly affluent urban population. The surrounding rural areas have been able to respond to these demands as a result of the greater freedom of economic action granted to rural leaseholders through the 1980 Household Responsibility Scheme. This has allowed small rural leaseholders to transition from the production of subsistence, mainly grain, crops to the supply of more profitable horticultural produce for the urban market and to transition from simple productivism by diversifying into rural tourism ventures such as farm shops and pick your own schemes, farm stays, restaurants using local produce and even rural heritage attractions such as folk villages.

The context of Kang's study is the conversion of locally oriented, primarily pastoral land in South-Eastern Turkey to intensive, export-oriented crop production. This change has been instigated by traditional owners of large estates who now have heavy demands for seasonal labour. These demands are being satisfied by large numbers of migrant workers without whom this agricultural transformation would have been impossible. However, this land use dynamism is occurring in some of the most conservative areas of Turkey in cultural and religious terms. Kang's chapter therefore focuses on the social and practical challenges faced by the female members of the extended family groups who make up the migrant labour force.

Kamuti's study of Zimbabwe is also concerned with the challenges contingent upon agricultural change though, in this case, the major challenge is environmental and the change is, if anything, the opposite of the "depeasantization" which characterizes the other three studies in this section. Land reforms since 1980, and particularly the "fast track" programme since 2000, have broken up the extensive holdings of white, settler colonial farmers and facilitated the development of smallholdings on both former settler and public land. One result of this has been a massive expansion in smallholder tobacco production. Smallholder methods of tobacco curing are extremely inefficient, with approximately 14 kilograms of wood being required to cure every kilogram of tobacco. Even in protected areas, large amounts of native forest are currently being cleared for the production and processing of this problematic and carcinogenic crop, a cost which needs to be balanced against any amelioration of rural poverty resulting from this development.

Part II: Recognizing Rural Demographic Diversity

This second sub-theme highlights the rapidly changing demographic landscape of rural regions. While demographic changes impact rural economies, both the rural economy and its representations in turn, shape demographic composition. This section invites the reader to explore the complexity of rural systems, including the interactions between levels of decision-making, within a geographical context combining local resources and connectivity.

Worldwide, rural areas experience important demographic changes which are linked to population growth, availability of land and water resources, connectivity to cities and the global world, diversification of the local economy, and finally international and regional migrations. Population growth and decline is one of the powerful driving forces that shape rural landscapes and societies. "Recognizing rural demographic diversity" is an essential step in understanding rural transformations. It involves the assessment of pressure on resources, the examination of adaptations and the study of the effect of out- and in-migration, including the questions of changing ethnic, racial, age and indigenous profiles.

In many places, we are witnessing a hybridization of life in the countryside which is intimately connected with the city (Pouzenc and Charlery de la Masselière 2020). The group, or the family, sends some of its members to the metropolis in order to earn money which is returned to the homeland. Communications technologies, such as mobile phones (Pasini 2020), change relations with the city and the global world and strengthen community bonding. In other parts of the world, urban populations move to the countryside to enjoy its amenities. Some of these new residents frequently commute to the city, others have opted for new ways of life. Some are retired. Others want to take a break or change their lifestyle to become rural people or to create a new generation of peasants looking for more natural agricultural practices (Van der Ploeg 2018).

Many rural areas, particularly those with natural amenities, struggle to manage the social and economic disparities created by second home ownership, tourism economies and seasonal work (Amit-Cohen 2013; Stedman 2006; Jones and Selwood 2013). Another group experiences gentrification, which pushes less wealthy people to move away and to abandon activities in the primary sector. In other places, long-term economic decline and out-migration create challenges including rapidly ageing populations and their increased need for health care options, rural poverty alleviation and the provision of services, such as schools, and of employment opportunities for youth and young adults (Long et al. 2013; Woods 2005). Further challenges arise when new industries relocate to rural places and bring with them new ethnically and socially diverse populations (Maher 2013; Barcus and Simmons 2013), creating demands for rural communities to provide education, housing and health care for new populations.

Rural populations will continue to evolve with ongoing transitions to knowledge-based economies and societies as the notion of distance changes and the need for spatial concentration weakens, and the focus on personal well-being reinforces the appeal of remote and green areas (Jousseaume 2020). Many rural areas are becoming sites of commodification where increasingly diverse representations of rurality coexist, which often leads to diverse forms of physical reality (Halfacree 1993) and to a proliferation of ways of dwelling (Schmitz 2000).

The chapters in this section describe local situations in the USA, Ireland, Poland and Burkina Faso where international influences have deeply transformed rural ways of life and society. The authors question especially changes in rural population composition which have become more diverse due to both international and local migration. In the Middle West of the USA, Barcus and Lanegran address the question of the ethnic diversification that has accompanied economic change and raise the issue of land availability for newcomers. "New residents mean new opportunities and challenges for communities" (Barcus and Lanegran, 2019, 16). Cawley's chapter analyses the last Irish census to show how remote rural areas may also welcome immigrant workers to respond to the lack of local workers for several industries, such as horticulture, meat processing and tourism. She identified several rural ethnic concentrations. In both studies, the arrival of foreigners changes the population composition and leads to the appearance of new services, including new forms of religion in the rural settlements.

In Poland, Dmochowska-Dudek and her colleagues reflect on the transformation of the local rural society and of the public spaces in villages. The decline in the number of farmers and the development of new activities, including rural tourism, open up both new prospects and perspectives for residents. The authors question the role of both local populations and external initiatives, using a neo-endogenous approach to rural revitalization. They analyse the perception of public space as a central issue and consider its evolution in connection with the new expectations for sociability in order to provide recommendations to improve the quality of public spaces in rural areas.

Finally, Moseley and Pessereau focus on the situation of women in Burkina Faso. Increasingly, women worked in the fields, while men sought remunerative activities, including gold mining elsewhere. The authors show how the New Green Revolution for Africa, which promotes the utilization of new agricultural inputs, including generic herbicides based on glyphosate, has replaced longstanding practices adapted to the local physical conditions and changed the gender distribution of labour. While this revolution has changed habits and saved time which can be given over to other activities, it has led to health problems that disproportionately affect the poorest women and their children.

Part III: Rural Innovations and Rural–Urban Partnerships

As the first two themes have illustrated, demographic change through migration, ageing and lifestyle preferences coupled with changing land uses and structures of production and consumption are yielding greater demand for the reimagining of rural life and rural landscapes. This reimagination is taking a multiplicity of forms. This sub-theme highlights conventional and perceptual barriers between rural and urban regions and examines activities and opportunities for building healthy and sustainable regional networks, including innovative and collaborative activities that span economic, social, environmental and administrative dimensions. We especially highlight opportunities gained from regional collaborations. These collaborations combine the complementing strengths and resources of rural and urban areas.

Ongoing processes of globalization have transformed the rural landscapes of the late twentieth and the early twenty-first century, including the relationships between rural residents, urban residents, land uses and activities. While change in rural places is not new to our contemporary time period, Woods notes that "[c]ontemporary rural change is … distinguished by two characteristics. These include the *pace and persistence* of change and the *totality and interconnectivity* of change" (Woods 2005, 30). He continues, arguing that "[r]ural areas … are tightly interconnected by global social and economic processes that cut across rural and urban space in a condition of advanced globalization" (Woods 2005, 30). Such transformation is evident at global, regional, national and more local scales and has created new flows of information, people, goods and services. While Roep and Wiskerke (2004) further frame the shift (after Marsden (2003)) as taking three forms, beginning with the agro-industrial model, post-productivist model and more recently the sustainable rural development model, Marsden and Van der Ploeg describe these transformations as "the *rural web*," which "… is a complex set of internally and externally generated interrelationships that shape the relative attractiveness of rural spaces, economically, socially, culturally and environmentally" (2008, vii). Thus, the conceptualization of rural areas as a complex and interconnected set of actors, inclusive of residents, industries, sectors, small businesses and governance processes embedded within broader territories, has challenged rural areas to reimagine their strategies for rural development. Key to success here is not only the role of facilitating structures, such as programmes that encourage collaboration, but also local-scale collaborative efforts and innovations.

Along with these reimagined flows are emerging governance structures that encourage new linkages across geographies. For example, beginning in the 1990s, new, more territorially based conceptualizations of space emerged in the European Union (EU). These new structures emphasized the importance of inter-linkages and relations within and across territories. So-called *neo-endogenous* structures have transformed how rural communities and economies are viewed and how they engage in rural development activities. The LEADER programme stands as an example in the EU. It is described as having a "… territorial, participatory and endogenous approach to rural development," which focuses on enabling residents to pursue development strategies that leverage local knowledge and resources (Bock 2012, 60) and act as mobilizing forces in "activating local stakeholders across all sectors" (Dax and Oedl-Wieser 2016, 30).

> The territorialization approach also emerges endogenously, as local autonomous initiatives. These are particularly evident in collaborative actions involving food producers or the tourism sector but also include regeneration initiatives of a more comprehensive nature…
>
> (Ray 2006, 281)

The use of local knowledge and resources, however, also finds its footing in ideas of social innovation. In this framing, rural residents work collaboratively to revitalize the rural economy based on collective actions and knowledge exchange (Bock 2012, 61). This new framing of rural development is transforming rural places. A key aspect of these transformations includes how relationships between and within territories and across urban–rural dimensions and partnerships are being redefined. Moving away from a purely sectoral model of rural development, communities are leveraging new branding and new businesses to create a more dynamic rural environment. At the micro-scale, this involves the reimagination of family businesses that might now emphasize regional or local branding to attract consumers or possibly tourists to their business. One example might be the farm-to-table movement, in which new linkages and relationships between farmers, restaurants and consumers are celebrated as a unique or special part of the consumption experience. Instead of exclusively working to minimize costs and maximize production, the development of new services and products, including local high-quality food, leisure space and nature conservation, become part of local development strategies (Roep and Wiskerke 2004). Urbanites may travel for day trips or holidays to experience such rural imaginations, while the farmers and restauranteurs find themselves selling not just agricultural products but the experience of local food.

Rural tourism exemplifies yet another dimension of a more territorially based conceptualization of rural places. Territories with distinct identities are perceived to attract more tourists. Establishing such identities and destinations requires individual businesses to work collaboratively and participatorily across sectors, inclusive of lodging, food and activity-based venues. Such collaborative work requires the leveraging of social capital and socio-economic resources to market a territory to consumers. The co-evolution of tourism and tourism networks may help diversify ideas and strategies (Brouder 2012), while collaboration and social learning complement efforts to revitalize rural places (Bock 2012).

Beyond the broader structures of territorialism that facilitate collaborative rural development strategies, are individuals within these areas who are, themselves, innovators. Brouder, for example, argues that there exists "latent social capital in rural places," particularly in peripheral rural places which may have formerly been resource dependent (2012, 384). Such latent social capital may emerge from groups previously marginalized within rural areas, inclusive of women (Ní Fhlatharta and Farrell 2017) and youth, or from in-comers who bring their own ideas and innovations.

This theme also attempts to provide insights into the rapidly changing landscapes and livelihoods of local rural producers and consumers. There are many forms of rural innovation and the literature reflects these many diversities. Taking a place-based approach helps illuminate linkages between innovations in each place and how these places and ideas are deeply connected to both processes of globalization and to more proximate urban areas. The chapters comprising this section of the edited volume focus specifically on the intersection of agriculture and food production in creating new or renewing old linkages between urban and rural places. Case studies come from Vietnam, Belgium, Morocco and Spain and emphasize linking local knowledge and people with global knowledge (Sanders 1994), recognizing opportunities, empowering people and building regional networks. Importantly, these case studies emphasize the growing interconnectedness and interdependencies of urban markets and rural food systems at a range of scales from proximate urban and rural areas to more global consumer-driven demand, particularly in light of new conceptualizations of food systems as needing to be "local" and "sustainable."

The theme of food production and consumption underscores new entrepreneurial endeavours in particular. Bousbaine and Schmitz, for example, write that

> Recently, in many places in the Global North, especially former industrial cities (such as Detroit, Manchester, Milan, and Liege), citizens and politicians reconsidered their links to food and their agricultural belts. This situation can lead to a profound rural transformation of the countryside and agricultural practices near urbanized areas as well as a renewed relationship and governance between towns and countryside.
>
> (Chapter 13, 217)

Their work highlights how a renewed interest in local food and regional agriculture in an urban setting has evolved into new relationships between farmers and urban residents and an "emergence of innovative agro-food systems, which constitute the belts of local food production around Liège and Charleroi, Belgium" (Chapter 13, 218). The authors focus on the synergies being created between urban and rural areas and how different perspectives on agricultural organization are emerging and changing the local framing of food and food production as urban residents seek creative ways to enhance and leverage local food systems for both sustainability and health purposes.

The second and third case studies presented in this section emphasize the role of rural transformations linked to agriculture, agritourism and cultural heritage. For example, Fatima Gebrati focuses on the transformation of rural space and the role of rural heritage for creating opportunities for marginal area revitalization in the High Atlas of Marrakesh. This chapter focuses on the relationships between local, social and economic vitality, national and global trends and the dynamic relations between hinterland and broader state structures. Gebrati's chapter focuses on analysing the issues related to the transformations of these marginal spaces by insisting more particularly on the nature of the valorization of the local specificities and the rural heritages.

Joan Tort-Donada and Jordi Fumadó-Llambrich take up the ideas of rural innovation and local development through the perspective of a single agricultural cooperative in Catalonia, Spain – the *Cooperativa Hortec*. The authors argue that by adopting

> ...a cooperative business structure – a structure that boasts a long tradition in Catalonia – *Hortec* explicitly defends innovative values such as agro-environmental sustainability and the adoption of an integrated production and marketing line – designed to "connect" small producers throughout Catalonia with centralized distribution centers, including Barcelona's fruit and vegetable market, *Mercabarna*.
>
> (Chapter 10, 169)

Our last case study comes from Vietnam, highlighting the role of globalization in the governance of food chains and the emerging emphasis on food safety within Global South countries. This chapter analyses the reorganization of supply chains towards producing safer agricultural products with a case study of safe vegetable production in Vietnam. Kim et al. highlight the importance of social networks and place-based knowledge in the evolution of food safety systems at multiple scales. Linking the demands of urban residents and consumers with global ideals in food safety places new constraints and opportunities on largely small, distributed rural producers.

Conclusions

In this volume, we invite you to explore the wide-ranging and multidimensional nature of rural transformations occurring across the globe. From our perspective, the rapid evolution of demographic and land use change processes, coupled with multi-scalar economic policies and multi-industry collaboration and new communication and other technologies are transforming rural social, economic and physical landscapes. We invite you to explore these transformations but also to appreciate the diversity of perspectives required to grasp these contemporary ruralities in the following chapters.

References

Amit-Cohen, I. 2013. Heritage landscape fabrics in the rural zone: an integrated approach to conservation. In *Globalization and New Challenges of Agricultural and Rural Systems: Proceedings of the 21st Colloquium of the CSRS of the IGU*, eds. K. Doo-Chul, A.-M. Firmino, and Y. Ichikawa, 79–88. Japan: Nagoya University Press.

Argent, N., Tonts, M., Jones, R., and Holmes, J. 2014. The amenity principle: internal migration and rural development in Australia. *Annals of the Association of American Geographers* 104: 305–318.

Barcus, H. R., and Lanegran D. A. 2019. Ethnic Restructuring, Land Use Change and New Farmers: The Case of Dakota County, Minnesota. In *Sustaining Rural Systems: Rural Vitality in an Era of Globalization and Economic Nationalism, Program*, ed. IGU-CSRS, 16. St Paul (Mn): Macalester College.

Barcus, H. R., and Simmons, L. 2013. Ethnic restructuring in rural America: migration and the changing faces of rural communities in the Great Plains. *Professional Geographer* 65 (1): 130–152.

Belich, J. 2009. *Replenishing the Earth: The Settler Revolution and the Rise of the Anglo-World 1783-1939*. Oxford: Oxford University Press.

Bock, B. B. 2012. Social innovation and sustainability: how to disentangle the buzzword and its application in the field of agriculture and rural development. *Studies in Agricultural Economics* 114: 57–63.

Brouder, P. 2012. Creative outposts: tourism's place in rural innovation. *Tourism Planning & Development* 9(4): 383–396.

Curry, G. N., Koczberski, G., and Selwood, J. 2001. Cashing out, cashing in: rural change on the South Coast of Western Australia. *Australian Geographer* 32: 108–124.

Dax, T., and Oedl-Wieser, T. 2016. Rural innovation activities as a means for changing development perspectives – an assessment of more than two decades of promoting LEADER initiatives across the European Union. *Studies in Agricultural Economics* 118: 30–37.

Gold, J. R. 1991. Fishing in muddy waters: communication media and the myth of the electronic cottage. In *Collapsing Space and Time: Geographic Aspects of Communication and Information*, eds. S. D. Brunn, and T. A. Leinbach, 327–341. London: Harper Collins.

Halfacree, K. H. 1993. Locality and social representation: space, discourse and alternative definitions of the rural. *Journal of Rural Studies* 9 (1): 23–37.

Holmes, J. 2006. Impulses towards a multifunctional transition in rural Australia. *Journal of Rural Studies* 22: 142–160.

Jacoby, K. 2014. *Crimes against Nature: Squatters, Poachers, Thieves and the Hidden History of American Conservation with a New Afterword*. Berkeley and Los Angeles, CA: California University Press.

Jones, R., and Diniz, A. M. A., eds. 2019. *Twentieth Century Land Settlement Schemes*. London and New York: Routledge.

Jones, R., and Selwood, J. 2013. The politics of sustainability and heritage in two Western Australian Coastal Shack communities. In *Globalization and New Challenges of Agricultural and Rural Systems: Proceedings of the 21st Colloquium of the CSRS of the IGU*, eds. K. Doo-Chul, A.-M. Firmino and Y. Ichikawa, 89–100. Japan: Nagoya University Press.

Jousseaume, V. 2020. Imaginer un urbanisme rural contemporain. In *Etudier les ruralités contemporaines*, eds. M. Pouzenc, and P. Charlery de la Masselière, 57–70. Toulouse: Presses Universitaires du Midi.

Long, H., Yurui, L., Yansui, L., and Xingna, Z. 2013. Population and settlement change in China's countryside: causes and consequences. In *The Sustainability of Rural Systems: Global and Local Challenges and Opportunities*, eds. M. Cawley, A. M. d. S. M. Bicalho, and L. Laurens, 123–133. Galway: Whitaker Institute.

Lowenthal, D. 1997. *The Heritage Crusade and the Spoils of History*. London: Viking.

Maher, G. 2013. Attitudes to Brazilian migrants in rural Ireland in conditions of economic growth and decline. In *The Sustainability of Rural Systems: Global and Local Challenges and Opportunities*, eds. M. Cawley, A. M. de S. M. Bicalho, and L. Laurens, 161–172. Galway: Whitaker Institute.

Marsden, T. 2003. *The Condition of Rural Sustainability*. Assen: Van Gorcum.

Marsden, T., and Van der Ploeg, J. D. 2008. Preface: exploring the rural web. In *Unfolding Webs, the Dynamics of Regional Rural Development*, eds. J. D. Van der Ploeg, and T. Marsden, vii–ix. Assen: van Gorcum.

Ní Fhlatharta, A. M., and Farrell, M. 2017. Unravelling the strands of 'patriarchy' in rural innovation: a study of female innovators and their contribution to rural Connemara. *Journal of Rural Studies* 54: 15–27.

Pasini, J. 2020. Le téléphone portable dans le Moungo central (Cameroun): des territoires du développement aux systèmes de ressources. In *Etudier les ruralités contemporaines*, eds. M. Pouzenc, and P. Charlery de la Masselière, 209–218. Toulouse: Presses universitaires du Midi.

Pequeno, R., and Elias, D. 2015. (Re)estruturação Urbana e Desigualdades Socioespaciais em Região e Cidade do Agronegócio. *GEOgraphia* 17: 10–35.

Pouzenc, M., and Charlery de la Masselière, P. 2020. *Etudier les ruralités contemporaines*. Toulouse: Presses Universitaires du Midi.

Ray, C. 2006. Neo-endogenous rural development in the EU. In *Handbook of Rural Studies*, eds. P. Cloke, T. Marsden, and P. H. Mooney, 278–291. London: Sage Publications.

Roche, M., and Argent, N. 2015. The fall and rise of agricultural productivism? An Antipodean viewpoint. *Progress in Human Geography* 39: 621–635.

Roep, D., and Wiskerke, J. S. C. 2004. Reflecting on novelty production and niche management in agriculture. In *Seeds of Transition*, eds. J. S. C. Wiskerke, and J. D. Van der Ploeg, 341–356. Assen: van Gorcum.

Sanders, S. R. 1994. *Staying Put: Making a Home in a Restless World*. Boston, MA: Beacon Press.

Schmitz, S. 2000. Modes d'habiter et sensibilités territoriales dans les campagnes belges. In *Des campagnes vivantes: un modèle pour l'Europe?*, ed. N. Croix, 627–632. Rennes: Presses universitaires de Rennes.

Stedman, R. C. 2006. Understanding place attachment among second home owners. *American Behavioural Scientist* 50 (2): 187–205.

van der Ploeg, J. 2018. *The New Peasantries: Rural Development in Times of Globalization*. Abingdon and New York: Routledge.

Wilson, G. 2001. From productivism to post-productivism … and back again? Exploring the (un) changed natural and mental landscapes of European Agriculture. *Transactions of the Institute of British Geographers* 26: 77–102.

Woods, M. 2005. *Rural Geography*. London: Sage Publications, Ltd.

Part I

Agricultural and Land Use Transitions

2 Agribusiness Towns, Globalization and Development in Rural Australia and Brazil

Michael Woods

Introduction

Across the world, traditional farming towns are facing the challenge of globalization in agriculture. Towns that for decades formed the hubs for dispersed farming communities as sites of markets, livestock marts and food processing plants, suppliers of farm inputs and labour and places where farmers met each other, banked their money, shopped for household goods and contributed to local society have become increasingly integrated into transnational networks with transformative effects. Scholarship in rural studies has tended to focus on the more dramatic, negative examples of farming districts that have struggled to adapt to globalization. Declining small towns with diminishing populations and depressed economies as local farms have been left behind by global competition and consolidation of agri-food production or places where agriculture has been eclipsed economically and culturally by diversification into new sectors, such as tourism. Less attention has been directed at rural towns in agricultural regions that have prospered with globalization, the emerging 'global farmlands' (McDonagh et al. 2015) that have become the engines of concentrated food production to supply global marketplaces.

Although globalization has been good for such towns economically, its social, cultural and environmental impact in the localities has been no less transformative than in declining towns. Moreover, a key element of transformation has commonly been the restructuring of local agriculture away from a system comprised of independent small farmers, cooperatives and local processors to a system dominated by transnational agribusiness. Family-run processing companies and farmers' cooperatives have been taken over by large multinational corporations, local short supply chains have been stretched and displaced by complex transnational networks, local inputs and products have been substituted by commercial imports, and land and farm operations have been bought by foreign-owned corporate farmers and investment funds. These changes have wider reverberations through rural communities and small towns, creating new dynamics of power in relation to local politics and decisions concerning issues of economic development, land use planning and infrastructure, and thus on the capacity of small towns to negotiate their own future paths.

DOI: 10.4324/9781003110095-3

In order to explore these dynamics and their consequences for small farming towns, this chapter draws on the concept of the agribusiness city (*cidade do agronegócio*) developed by Brazilian geographers Denise Elias and Renato Pequeno (Elias and Pequeno 2007; Pequeno and Elias 2015). As detailed further below, an agribusiness city is an urban centre that is oriented towards the supply and support of corporate agriculture such that the interests of agribusiness dominate the local political economy, housing, land and labour markets. The label has been particularly applied to expanding cities in central Brazil that have boomed with the globalization-driven development of the agricultural frontier. This chapter modifies the concept of the agribusiness city in two ways. Firstly, by rearticulating the concept slightly as 'agribusiness towns' in order to apply its principles to smaller towns where farming has always been the mainstay of the economy and secondly, by examining whether the concept can be translated to geographical contexts outside Brazil.

As such, the chapter presents a comparative analysis of two towns, Dom Pedrito in the state of Rio Grande do Sul in Brazil and Smithton, in Tasmania in Australia. The two towns are both situated in sparsely populated rural districts in peripheral regions of their country as well as at a similar latitude. They are both products of settler colonialism and are located in regions that have prospered with agricultural globalization, accompanied by a growing presence of agribusiness, and, in both localities, there has been a recent reorientation of farming towards commodities targeted at export markets. Yet, the towns sit in different political–economic contexts that mediate the impacts of globalization and the entry of transnational business, and which shape the parameters for future development.

The chapter therefore examines the following questions: How have the agricultural economies of the case study localities been restructured through globalization and what role has transnational agribusiness played in this transformation? What is the visible presence of agribusiness in the towns and what impact does it have on the geographies of the area? What influence does agriculture have in the social, political, economic and land use structures and systems of the towns? What are the potential future trajectories for the towns and how are these enabled or constrained by the interests of agribusiness?

The research on Dom Pedrito and Smithton forms part of a larger project investigating how globalization has been reproduced and contested through rural localities in the emergent global countryside (Woods 2007), which has involved 32 case studies in 15 countries, each selected to illustrate specific aspects of globalization.[1] Dom Pedrito was selected as an example of export-oriented agricultural transition in an emerging economy and Smithton initially as a case study of international land investment. Fieldwork in Dom Pedrito was conducted in March 2018, involving 18 interviews with farmers, agricultural advisors, local politicians and other local stakeholders and commentators, as well as field observation, supplemented by later collation and analysis of statistical data and secondary documents primarily sourced online.[2] Fieldwork in Smithton was conducted in May and June 2018, with field observation and nine interviews with farmers, agricultural advisors, politicians and other stakeholders in Smithton and the wider region supplementing more extensive analysis of secondary sources including newspaper articles, state and local government papers, research reports, local histories, publicity brochures, maps, photographs and statistics, sourced from local libraries in Smithton and Burnie, the State Library of Tasmania, the University of Tasmania and online archives.[3]

Agribusiness Towns in Brazil and Australia

An important feature of the globalization of agriculture has been the corporatization of agricultural systems and the propagation of an agribusiness ethos (Robinson and Carson 2015). Although definitions of agribusiness vary, as a collective noun it can be understood as referring to the operation of agriculture on commercial lines, following generic business and management principles with significant corporate involvement in multiple elements of agri-food value chains. As such, agribusiness may be contrasted to subsistence and peasant agriculture, as well as small-scale family farming, even where such farming has commercial dimensions through market exchanges but includes hybrid forms of 'family farm entrepreneurs' who combine family ownership and labour with agribusiness practices (Pritchard et al. 2007; Cheshire and Woods 2013). Additionally, as a singular noun 'agribusiness' can be applied to individual businesses active in the agri-food sector, including not only corporate farmers but also seed companies, agri-chemical manufacturers, machinery suppliers, contract farm labour and service providers, crop processing firms and so on.

It is these supporting agribusinesses that are the facilitators of agricultural globalization, connecting farms that are necessarily fixed in place with transnational supply chains and markets. As agribusinesses conform to conventional logics of business operation, such as economies of scale, agricultural globalization has been characterized by corporate concentration – with global markets for inputs such as seeds and fertilizers dominated by a few transnational firms – and vertical integration, with corporate alliances linking components of agri-food production chains (Hendricksen and Heffernan 2002; Clapp 2018). Furthermore, as profit maximization by agribusinesses demands constant innovation and staying ahead of competitors, their farmer clients are put under pressure to intensify, enlarge and adopt new technologies and business practices. Accordingly, global agribusiness is associated not only with depeasantization and the decline of family farming but also with farm consolidation, specialization, automation and the development of digital agriculture. Responding to these pressures requires capital input, thus contributing to the financialization of agriculture, including speculative investment in farmland by banks, pension funds and asset managers (Lawrence et al. 2015; Clapp and Isakson 2018).

In common with other economic sectors, global agribusiness has become a footloose industry, with just-in-time production for liberalized transnational markets concentrated in the most efficient locations. Favoured regions have become 'global farmlands' (McDonagh et al. 2015), producing and exporting commodities around the world, whilst farms outside these regions, many in traditional farming districts, have faced falling incomes as they struggle to compete, creating a differentiated global countryside (Woods 2007; Woods 2014). The decline in the economic and political significance of farming in many rural regions across the Global North is therefore counterpoised by the 'global farmlands', in which agribusiness continues to occupy a central economic position and to exert influence not only in rural communities but also in the towns and cities that service them.

It is in this context that Elias and Pequeno (2007) introduce the concept of the 'agribusiness city' (*cidade do agronegócio*) to describe small- and medium-sized regional centres that are closely connected to the expansion of global agribusiness in adjacent rural areas. Agribusiness cities attract companies that establish branches and offices to service globalized agricultural production, but more than that an agribusiness city 'is one whose functions to meet the demands of globalized agribusiness are hegemonic over other functions' (Elias and Pequeno 2007, 30 (translated)). As such, agribusiness cities have governance regimes that act to reproduce global agricultural capital, shaping policies on land use, infrastructure, economic development, housing and taxation. These can include prioritizing agribusiness developments in prime sites along highways but also consumption-focused developments targeted at income earned from agribusiness, such as upscale housing that can often involve the displacement of lower-income workers. In such ways, Elias and Pequeno note, agribusiness cities start to replicate the socio-spatial inequalities and urban problems of larger, established cities.

Many of the examples of agribusiness cities cited by Elias and Pequeno and other scholars are located in the interior regions of Brazil where the expansion of globalized agribusiness with export-oriented crops of soy, wheat and fruit has involved extensive land transformation (including deforestation), substantial in-migration and rapid urbanization (Elias and Pequeno 2007; Frederico 2011; Ioris 2018). Pequeno and Elias (2015), for instance, discuss the case of Luís Eduardo Magalhães, an agribusiness-dominated city in the rich agricultural MATOPIBA region, that expanded from a population of 10,000 people in 2000 to over 76,000 in 2014, having been formed as an independent municipality in 2000 through the machinations of agribusiness interests.

However, Elias and Pequeno (2007) also indicate that features of agribusiness cities can also be found in smaller and long-established settlements that have experienced the entry of transnational agribusinesses, 'disrupting the previous socio-spatial formation [and] bringing new territorial dynamics and political and socio-cultural conditions' (Elias and Pequeno 2007, 30 (translated)). This chapter focuses on these smaller places, characterized here as 'agribusiness towns', to examine the socio-economic and political implications of agricultural globalization for the localities and the influence of agribusiness interests through a case study of Dom Pedrito in Rio Grande do Sul, southern Brazil. Furthermore, noting that empirical elaboration of the 'agribusiness cities' concept has so far been limited to cases in Brazil, the chapter also explores whether the concept can be transposed to other 'global farmlands', such as Australia, with a second case study of Smithton in Tasmania. The two case studies are outlined below, with comparative discussion in the subsequent section.

Dom Pedrito, Rio Grande do Sul

Located in the far south of Brazil, on the border with Uruguay, the municipality of Dom Pedrito and its population of 40,000 people has always been a farming community. Historically, the local economy was based on livestock ranching, with cattle and sheep raised on the extensive properties that ranged across the undulating natural landscape of the pampa biome producing meat, leather and wool for domestic and export markets. In between the large estates, *pecuárista familiar* – small livestock farmers – eked out pluri-active livelihoods on small plots of land. The export trade in wool, referred to locally as 'white gold' (Farmer, interview DP18a), brought prosperity to the town and local society and place-identity reflected and celebrated the *gaucho* culture of the ranchers.

In the 1970s, falling world prices for wool precipitated by the development of synthetic fabrics and market-distorting actions of Australian woolgrowers prompted a switch away from sheep farming with landowners renting out sub-divisions. The tenancies were mainly taken up by German and Italian-descendent farmers from the centre of Rio Grande do Sul state, who introduced new crops of rice and soybeans and with them a sharper business mentality and more engagement with agribusiness networks. Traditional livestock ranching around Dom Pedrito had been considered unsophisticated and lacking in modernization. Animals were grass-fed and taken for slaughter to the abattoir in Dom Pedrito. The growth of arable farming created new demands and interacted with more complex supply chains. Processing plants for rice were built in the town and businesses supplying agri-chemicals and machinery opened.

The major impetus for agribusiness investment, however, came from the expansion of soy. Although there had been experiments with soybean cultivation in the district at the start of the twentieth century, the clay-heavy, ill-drained soils around Dom Pedrito were not well suited to growing soy, and the municipality was a relative latecomer to the modern Brazilian soybean boom. It was not until several years into the 2000s that the inflation of world soybean prices by demand from China combined with technological advances made growing soy in Dom Pedrito a gainful venture. Between 2010 and 2017, the area of the municipality planted with soy increased fivefold from 20,000 hectares to 100,000 hectares, largely at the expense of livestock farming, which halved in extent (Figure 2.1).

In 2018, Dom Pedrito produced 192,000 tonnes of soybeans, 88% of which were exported to China, with smaller volumes going to South Korea, Spain and five other countries (TRASE 2020), valued at around 200 million Brazilian Reais (approximately US$ 60 million). Overall, 64% of the GDP of Dom Pedrito is derived from agribusiness and 85% of the municipality's tax revenue comes from agribusiness (Interviews DP6, DP15). The fiscal contribution of agribusiness flows not only from commodity exports but also from the involvement of agribusiness companies and services in each stage of the production chain. Indeed, the precarity of soy cultivation in Dom Pedrito deepens the dependency on agribusiness, far more so than for rice or livestock farming.

In order to overcome the environmental disadvantages of the local soil, growing soy in Dom Pedrito requires modified seeds sourced from transnational

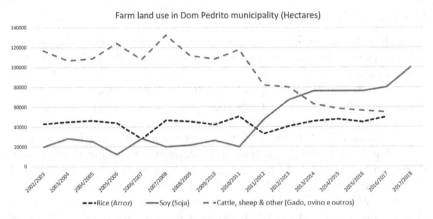

Figure 2.1 Change in farmland use in Dom Pedrito municipality, 2002–2018.
Source: EMATER.

biotechnology firms, agri-chemical inputs from multinational corporations, irrigation systems constructed by US-based companies and tractors and equipment imported from international manufacturers. It also requires finance. As soil conditions around Dom Pedrito are classified by the state agricultural agency as poor for soy cultivation, farmers find it difficult to access finance for growing soy from commercial banks and thus rely on agribusiness firms themselves for credit, as one farmer explained:

> [Capital] comes from the bank, only that the bank is very restrictive, has little appeal and very large demands. They want many guarantees, real estate and things, so sometimes you don't have it right. Sometimes you already have a guarantee in property for a machine or something, so you have no way to access the bank's credit. So you use the industry … [It's] more expensive, much more expensive.
>
> (Farmer, interview DP12)

The financing from agribusiness companies normally comes in the form of a package that provides seeds and inputs upfront in return for the sale of harvested soybeans at a pre-agreed price:

> The [farmer] says: I have 200 hectares – an example – and then the company says for 25 sacks per hectare I provide you seed, fertilizer, herbicide, fungicide, technical assistance, 25 sacks is mine per hectare, and the rest is yours.
>
> (Agricultural advisor, interview DP1)

The package is usually linked to a transnational seed producer, such as Bayer or BASF, with other inputs supplied by associated companies. Technical assistance is commonly included, with advice on crop management provided by the cor-poration's representatives that further works to benefit its capital accumula-tion. Each element of the package ties the farmer more closely to globalized agribusiness.

The tightening of the relationship between local farmers and transnational agribusiness has been manifested in Dom Pedrito with a proliferation of inter-mediary businesses that act as agents for transnational firms, including seed merchants, agri-chemical and fertilizer suppliers and equipment dealers, as well as providers of agribusiness services such as irrigation systems, aerial spraying, agri-technology support, transport, insurance, financial services and agricultural consultants. Many of these are located on the northern edge of the town, along the main highway, forming an agribusiness fringe that has extended the urban area.

As observed by Elias and Pequeno (2007) for agribusiness cities elsewhere in Brazil, the consolidation of transnational agribusiness in Dom Pedrito has also intensified local dynamics of the urbanization of the population. Soy culti-vation is less labour intensive than livestock ranching, requiring fewer perma-nent on-farm employees. Consequently, there has been a movement of former farmworkers into the town and a restructuring of the agricultural labour market around short-term contracting. Similarly, a number of smallholders (*pecuárista familiar*) have moved to the town as they have been squeezed off their land by a combination of rising costs of inputs, competition for land, diminished local mar-kets for produce and the impacts of pollution from aerial spraying and incursion of invasive species linked to soy farming. As the rural population has decreased, the municipality has controversially started to close rural schools, further accel-erating migration into the town as families seek to avoid journeys to school that could take two to three hours each way in the wet season. From an already high rate, the concentration of the municipality's population in the town of Dom Pedrito itself has increased to over 90%, leaving the extensive rural hinterland free for agribusiness production.

These developments have been actively supported by local governments as the drivers of economic growth. The estimated GDP per head for Dom Pedrito municipality increased from 5,590.45 Reais (approximately US$1,000) in 2000 to 10,970.59 Reais (US$2,000) in 2007 (at the start of the soy boom) and to 29,596.75 Reais (US$5,500) in 2015 (DeepAsk 2015) – significantly above the mean for its population band and making Dom Pedrito one of the richest munic-ipalities in the state (Interview DP15). Increased prosperity has encouraged new opportunities for consumption, with the opening of a Walmart supermarket in 2018 regarded locally as a milestone event. New houses have been built larger in size, and a university campus has been opened with a degree in agribusiness as its main offering. However, some in the town question who really benefits from the new wealth. Unlike meat and rice, harvested soybeans are transported directly by truck to the port at Rio Grande and exported unprocessed such that there is no value added to the product locally. Indeed, the food processing sector in Dom Pedrito not only misses out on soybeans but has been indirectly weakened by the expansion of soy, with the conversion of land away from livestock contributing to the closure of the town's abattoir in 2015 with the loss of 70 jobs. Neither are farmers necessarily beneficiaries:

> [The farmer and the agribusiness company have] an interrelationship that generates some employment, but the product after harvesting generates income for someone else [not the farmer], who is the trucker that carries it. Because it does not have [local] processing or industrialization in the production itself it generates some work, but it is not much.
>
> (Agricultural advisor, Interview DP1)

The reinvention of Dom Pedrito as an agribusiness town has therefore generated new fissures of inequality. Landowners, hauliers and the providers of agribusiness services have prospered but less so manual and semi-skilled workers, with displaced farmworkers not necessarily finding new employment. There are concerns especially about opportunities for young people and a rise in youth out-migration. Inequalities between farmers have also expanded, reflecting differential capacity to withstand market volatility. An intrinsic feature of the agribusiness model applied in Dom Pedrito is that risk is loaded onto the farmer who acquires debt against pre-agreed supply of produce to agribusiness buyers. The vulnerability of the farmer in this arrangement had been exposed at the time of our fieldwork in Dom Pedrito by the effects of drought and a fall in Brazilian rice prices due to competition from Paraguayan imports, compromising the ability of some farmers to meet commitments to agribusiness corporations or to service debt. As one local observer noted,

> The concept of agribusiness is the idea of the commodity. You produce a commodity, you produce to the market, you sell your production three or four years in advance. You spend all the money and then you have no money to pay when you need a loan, but that's agribusiness.
>
> (UNIPAMPA professor, Interview DP7)

Smithton, Tasmania

The farming centre of Smithton, with its population of 4,000 people, lies in the far north-west of the island state of Tasmania and is the largest settlement in the district of Circular Head (population 8,000). The region was colonized in the early nineteenth century by the Van Diemen's Land Company (VDLC), a British Royal Charter company that established its headquarters at Circular Head, 18 kilometres north-east of present-day Smithton, and an expansive 22,000-hectare farming property at Woolnorth, to the west. The swamp and forest between the VDLC holdings were gradually cleared and settled by pioneer European farmers – a process that continued up to the 1960s – and by 1900 a service centre had started to coalesce around Smithton on the Duck River (Pink and Ebdon 1988). The early farmers mainly grew vegetables for sale in Melbourne, across the Bass Strait, and supplied milk to the Duck River Co-op Butter

Factory in Smithton, a farmer-owned cooperative. During the Second World War, the Australian Government built a vegetable dehydration factory, later converted to a vegetable cannery, which brought 'manufacturing' to the town.

Through the twentieth century, the main industry of Smithton was timber, with sawmills supplied initially by the clearance of natural forests for farm settlements and later by tree plantations. However, political wrangling over forestry and conservation in Tasmania and an eventual 'peace agreement' that restricted logging of old-growth forests severely affected the industry in Smithton, with the largest sawmill closing in 2012. Media reports at the time presented north-west Tasmania as a 'troubled' and 'jobs-hit' region and, in response, both local politicians and outside commentators pointed to agribusiness, and especially the dairy sector, as the solution (ABC Rural News 2011; Evans 2012; Bird 2014).

Indeed, the local economy of the Circular Head district had already started to move towards the dominance of dairy farming. In the 2006 Census, dairy farming was the largest single industry in Circular Head, employing 452 people (or 12% of the workforce). This figure remained constant in the 2011 Census, at 451 people, but increased in 2016 to 524 people working in dairy farming (14.8% of the workforce), including 118 people living in the town of Smithton itself. Also, by 2016, over 40% of the dairy cows in Tasmania were being farmed in the Circular Head district (Newman et al. 2016).

Transnational investment by both companies and individuals had played a major part in driving the expansion of dairy farming in Circular Head. Dairy farmers from New Zealand had been moving into the district since the start of the century, attracted by cheaper land prices and suitable environmental conditions, assisted by reciprocal settlement and employment rights, and reassured by the familiarity of dealing with the New Zealand dairy giant Fonterra, which had opened a processing plant in nearby Wynyard. One local politician estimated that two out of five dairy farmers in Circular Head in 2018 were New Zealanders (Interview TAS1), but there were also new dairy farmers from Britain and the Netherlands. Other investment in Tasmanian dairying had come from agribusinesses, with one observer recalling,

> multiple farms [that] were sold to a Thai based company and other groups that were recently sold to Americans and managed by an Australian arm, then another group to Canadian money, sort of managed by New Zealand, sort of operating it'.
>
> (Industry stakeholder, Interview TAS4)

As a Circular Head farmer noted, agribusiness investment had more transformative capacity than individual farmers:

> I think mainly with the bigger companies coming in, they're bringing capital in. As much as everyone likes to keep it down at a smaller level type thing like that, often the individual farmer can't ... a small farmer can't do a big development of something like that. Whereas some of the bigger [investors] – and I suppose it's like with many corporate farmers coming in they're often

syndicates – they have superannuation funds or that sort of thing. At farmer level there's a little bit of angst about that, but [corporate investors] are the ones who are bringing it down. They're the ones who can bring in cattle to build something substantial.

(Farmer, Interview TAS3)

The largest and most significant dairy conversion project was on the VDLC property at Woolnorth, the oldest corporate farm in the state, and was funded by foreign capital. For most of its 180-year history, Woolnorth had been farmed for sheep and beef cattle and throughout that time it had been owned by the VDLC, though the VDLC itself had passed between several private owners. During the 1980s, new Italian owners had attempted to modernize the sheep-farming operations by introducing new breeds and investing in a new shearing shed, achieving profit for a while but fighting against the global trend of falling wool prices (Pink 2003). In 1993, the VDLC was acquired by New Zealand-based Tasman Agriculture (later bought by the investment fund of New Plymouth Council), who started to convert Woolnorth into the largest dairy farm in the southern hemisphere. The project involved creating semi-autonomous dairy operations, some converted from beef and sheep grazing, others carved from clearance of bush, each with a multi-million Australian dollar cost. By 2016, the VDLC operation comprised 25 dairy units over 8,272 hectares of land, milking 18,290 cows and producing nearly 100 million litres of milk a year (The Mercury 2016).

The VLDC was sold again in 2016, controversially, to a Chinese-owned company, Moon Lake Investments, that aimed to continue the dairy expansion with plans to build its own processing plant and to export fresh liquid milk directly to China (Figure 2.2). The latter initiative commenced in 2017, with around 10%

Figure 2.2 The VLDC Woolnorth property near Smithton, acquired by Moon Lake Investments in 2016.

Source: Author.

of Woolnorth's milk production being transported within 48 hours via Hobart and Melbourne airports and sold fresh by retailers in Beijing, Ningbo and Shanghai at a premium price of 77–79 RMB (US$11) per litre.

Initially, the growth of dairy farming outpaced local processing capacity and as such the development of food processing industries was adopted as a key priority in regional economic development. A new AUS$60 million (US$59 million) milk powder plant was constructed on a former sawmill site with government assistance by Australian dairy company Murray Goulburn and Japanese conglomerate Mitsubishi. The plant, acquired in 2018 by Canadian agri-food firm Saputo, joined a smaller UHT milk processing plant at Edith Creek, 14 kilometres south of Smithton (also owned by Murray Goulburn), an abattoir (owned by Melbourne-based Greenham's and exporting beef to East Asia) and the vegetable-processing factory (by now owned by Canadian firm McCain, who had added a French fries production plant) in forming a food manufacturing sector that was the largest employer in Smithton itself, collectively employing 227 people in 2016.

Additionally, the presence of large agribusiness corporations in Smithton created work for local businesses, including 'electricians and plumbers and builders and concreters' (Interview TAS3), whilst the expansion of dairying supported an array of agricultural services firms, including irrigation system installers, feed merchants and equipment dealers. As in Dom Pedrito, these concentrated in an 'agribusiness fringe', occupying sites on the edge of Smithton close to the Bass Highway and along a former railway line.

The development of agribusiness and food processing in Smithton was explicitly promoted by local government, both through local land use policies and through lobbying of state and federal governments for funding and favourable policies. State-sponsored, agribusiness-friendly projects included support for irrigation, enhanced electricity supply and one of Australia's first fibre-optic networks, as well as the opening of an agricultural training college in Smithton, intended to meet the demands of the beef and dairy industries (Sharp 2010; Acheson 2015; Mallinson 2015). For local politicians, cultivating agribusiness was an essential part of local government's function:

> We spend a lot of time talking to our [food] manufacturers in our area and that never used to happen a number of years ago … Not everybody in the community agrees with it, but when I first got on the council we were really rubbish [collection] and very little else. Now the community services and working with the industries is getting more and more important and I think government should probably do more than they do with it, but we do get good support from our government and they certainly work with us, but I think a lot of it or most of it is driven by councils. That's number one.
>
> (Local Government Official, Interview TAS1)

Yet, the strategy is also driven by awareness that the peripheral location of Smithton imposes limitations that could make it vulnerable to the footloose decision-making of transnational agribusiness. The vulnerability was exposed in

2010 when McCain closed the vegetable processing side of its factory in Smithton and moved production to New Zealand because supplies of vegetables had been affected by the conversion of land to dairy farming. Seven years later, in 2017, Murray Goulburn closed its plant at Edith Creek as part of a global restructuring of assets, with the loss of 115 jobs. The factory was bought by a Thai firm, Dutch Mill, which planned to produce UHT milk for export to Thailand, but by the time the plant reopened after a short hiatus, 93 of the redundant workers had already found new jobs, leaving Dutch Mill with a labour supply problem, whilst supplier farms had switched to different processors (Tasmanian Government 2018). Accordingly, issues of land and labour supply continue to be preoccupations for local politicians in Smithton, but they have only limited capacity to act with respect to land regulations and immigration laws set by state and federal governments.

Discussion

Dom Pedrito and Smithton have both transitioned from being traditional farming (and forestry) towns to become agribusiness hubs integrated into global production networks. The transformation has had substantial economic, social and environmental impacts in both localities. The restructuring of local agriculture around soybeans and dairy, respectively, as export-oriented commodities and inward investment by agribusiness firms have fuelled economic growth in both places, creating new jobs, raising incomes and promoting the proliferation of small businesses providing agribusiness services. These changes have had a physical manifestation, in the altered appearance of the farm landscape and in the development of 'agribusiness fringes' to both towns occupied by agribusiness outlets, processing plants and storage units. Increased prosperity has been reflected in new housing, shops, bars, restaurants and leisure opportunities in both towns.

As smaller towns, Dom Pedrito and Smithton have so far avoided the development of conventional 'urban problems', such as the congestion and crime observed by Elias and Pequeno (2007) in larger agribusiness cities. Yet, beneath the surface patterns of widening inequalities resonate with the agribusiness city model. The agribusiness boom has benefited some in the towns more than others, with lower-grade workers on farms and in processing plants taking the hit from modernization, corporate rationalization and the squeezing out of established local businesses. In Dom Pedrito, this has led to high unemployment, fierce competition for vacancies and a third of the population living on half the minimum wage (Trojahn 2020). In contrast, there is in effect full employment in Smithton, but many jobs are precarious and new inequalities have been created by agribusinesses filling labour gaps with migrant workers from Malaysia, South Korea, Thailand and Russia, who are tied to companies and have limited residence and employment rights (Interview TAS1).

Moreover, there are environmental impacts associated with the expansion of agribusiness. In Dom Pedrito, these are linked to the spread of soy monoculture, pressures on water supply from irrigation and the excessive use of agrichemicals.

Figure 2.3 Soybean fields encroaching on cattle pasture near Dom Pedrito.
Source: Author.

Around Smithton, environmental concerns arise from the clearance of bush for dairy conversions and risks of pollution from dairy operations. The landscape in both localities has been modified by the rise of soy cultivation and dairying, respectively, perhaps irrevocably. Fields of soybean encroach on the natural 'native field' (*campo nativo*) grassland around Dom Pedrito, replacing colours of light green and dusky brown with dark green (interspersed with the shimmering yellow of ripe rice) (Figure 2.3). The pasture is not easily restored after soy has been planted, as one farmer showed us, pointing out a field pocked with thistles and other invasive species inadvertently introduced with the soy. In Tasmania, native bush and vegetable fields have been converted into dairy pastures, although farmers complain about the conservation laws that prevent them clearing tree plantations. In both places, the refashioning of the landscape around agribusiness is further indicated by the proliferation of the paraphernalia of modern farming: irrigation systems, electric fences and power cables; concrete tracks and new miking sheds around Smithton (Figure 2.4); glistening new silos sprouting in the farmlands of Dom Pedrito. It is in these ways that 'landscape of the global countryside is inscribed with the marks of globalization' (Woods 2007, 493).

The hegemony of the agribusiness function in the towns is articulated in the resolute alignment of local government with agribusiness interests in spite of their mixed impacts. This has been demonstrated by seminal issues in both Dom Pedrito and Smithton in recent years. The agribusiness hegemony in Smithton was evident in local backing for Moon Lake's controversial acquisition of the VDLC in 2016. A rival bid was mounted by an Australian consortium led by a prominent Tasmanian entrepreneur and supported by mainland media and politicians concerned at growing Chinese influence. However, for local politicians

Figure 2.4 The dairy farming landscape near Smithton.
Source: Author.

and business groups in Smithton, it was the Australian-led bid that presented the greatest threat to 'progress' because of the entrepreneur's pro-environmental record:

> It was basically like having a vegan run an abattoir. [The entrepreneur] was not going to grow the business ... so that all the farmers would then go back to organic farming and we're going to put more back into planting trees. We could actually see that for the district she was not doing growth. She was going to be doing the opposite. There would be less cows out there which would mean employing less families. Whereas Moon Lake Investments they wanted to do the opposite. They wanted to grow the business.
>
> (Business group leader, Interview TAS3)

Similar narratives were articulated in debates over agrichemical use in Dom Pedrito. Commercially, competitive soy cultivation depended on genetically modified crops and the intensive use of agrichemicals, including the Roundup herbicide, usually applied by aerial spraying. Drift from sprayed agrichemicals has been linked to the collapse of local horticulture and bee-keeping activities, pollution of the Santa Maria River and high incidences of cancer. Yet, local political and agribusiness leaders have played down concerns. As one local critic contends, 'because the municipality is a supporter of the system, the public manager who manages it is to maintain the system, so he collaborates with the system' (Local campaigner, Interview DP2).

The pro-agribusiness regimes of Dom Pedrito and Smithton are influenced by tax revenues and informed by settler colonial discourses of 'progress', but they also reflect calculations that structural constraints (marginal conditions for soy cultivation in Dom Pedrito; restricted supply of land and labour in Smithton) mean that the presence of transnational agribusiness is contingent and tenuous.

The localities are exposed to decisions made in distant corporate headquarters, trends in major markets, such as China, and geopolitical interventions. They are also linked through globalized networks to other agribusiness towns and cities in other 'global farmlands'. What happens in Dom Pedrito and Smithton will have implications in these other places and vice versa.

Conclusion

In the differentiated global countryside, the increasing concentration of export-oriented commodity agricultural production in favoured regions that represent the most efficient sites of production for transnational agribusiness at any given time has been accompanied by the strengthening of the hegemony of agribusiness in these regions. This agribusiness hegemony is manifested in these localities not only by the primacy of agribusiness in farming systems and its substantial contribution to the local economy but also by the alignment of political elites with agribusiness interests, producing social, economic and environmental effects. Elias and Pequeno (2007) have captured these dynamics in the concept of the 'agribusiness city', applied to the emergence of new urban forms in the Brazilian interior whose function is to support agribusiness accumulation. The case studies of Dom Pedrito and Smithton discussed in this paper show that the broad concept can be adapted and translated both to smaller traditional farming towns undergoing transformation by agribusiness and to contexts outside Brazil.

Applying the 'agribusiness town' framework helps to reveal the processes through which rural localities are enrolled in global production networks and the assemblages of transnational agribusiness, how these processes produce both positive and negative impacts for places, with economic growth but also increasing inequality and environmental damage, and how they rely on the acquiescence of local governance. In cases such as Dom Pedrito and Smithton, it also highlights the contingency and precarity of their prosperity in competitive global markets; yet equally, as agency in the global countryside is distributed and relational (Woods 2007), the power of agribusiness is not unconstrained. Thus, whilst the concept of agribusiness cities or towns describes a specific urban form, they are not identikit places. The interaction of transnational agribusiness and rural place is mediated by local, regional and national actors, and indeed it is these differences that are exploited in the footloose strategies of global agribusinesses.

Acknowledgements

1. GLOBAL–RURAL, funded by a European Research Council Advanced Grant (No. 339567), 2014–2019. See http://www.global-rural.org for more information.
2. Research in Dom Pedrito was undertaken with the support of the Universidade Federal do Rio Grande do Sul and involved the author, post-doc researcher Francesca Fois and local research assistants Juliana

Moreira Gomes and Rodrigo Giesler Maciel. Interviews were conducted in Portuguese and transcribed and translated. Valuable guidance and advice were additionally provided by Alessandra Matte and Sergio Schneider.
3. Research in Smithton was undertaken solely by the author, supported by visiting scholar facilities and access provided by the University of Tasmania.

References

ABC Rural News. 2011. "Dairy an Option for Job-hit Northern Tasmania." *ABC Rural News*, January 6, 2011. EBSCO Australia/New Zealand Reference Centre Plus.

Acheson, M. 2015. Powering Up Dairy Industry. *Australian Financial Review*, February 21, 2015. EBSCO Australia/New Zealand Reference Centre Plus.

Bird, I. 2014. Circular Head on Way to Recovery. *Australian Financial Review*, May 25, 2014. EBSCO Australia/New Zealand Reference Centre Plus.

Cheshire, L., and M. Woods. 2013. Globally Engaged Farmers as Transnational Actors: Navigating the Landscape of Agri-food Globalization. *Geoforum* 44: 232–242.

Clapp, J. 2018. Mega-mergers on the Menu: Corporate Concentration and the Politics of Sustainability in the Global Food System. *Global Environmental Politics* 18: 12–33.

Clapp, J., and S. R. Isakson. 2018. *Speculative Harvests*. Black Point, NS: Fernwood.

DeepAsk. 2015. Confira do Produto Interno Bruto. Accessed July 23, 2020. http://www. deepask.com/goes?page=Confira-o-Produto-Interno-Bruto---PIB-por-tamanho-de-municipio-do-Brasil

Elias, D., and R. Pequeno. 2007. Desigualdades Socioespaciais: Nas Cidades do Agronegócio. *R. B. Estudos Urbanos e Regionais* 9, (1): 25–39.

Evans, M. 2012. We Need to Talk about the Northwest. *Griffith Review* 39: 239–246.

Frederico, S. 2011. As Cidades do Agronegócio na Fronteira Agrícola Moderna Brasileira. *Caderno Prudentino de Geografia* 33, (1): 5–23.

Hendricksen, M., and W. Heffernan. 2002. Opening Spaces through Relocalization: Locating Potential Resistance in the Weaknesses of the Global Food System. *Sociologia Ruralis* 42: 327–369.

Ioris, A. 2018. Place-Making at the Frontier of Brazilian Agribusiness. *GeoJournal* 83: 61–72.

Lawrence, G., S. R. Sippel, and D. Burch. 2015. The Financialisation of Food and Farming. In *Handbook on the Globalisation of Agriculture*, eds. G. M. Robinson and D. A. Carson, 309–327. Cheltenham, UK: Edward Elgar.

Mallinson, J. 2015. Agritas Graduates Key to Future Growth. *Australian Financial Review*, April 8, 2015. EBSCO Australia/New Zealand Reference Centre Plus.

McDonagh, J., B. Nienaber, and M. Woods, eds. 2015. *Globalization and Europe's Rural Regions*. Aldershot, UK: Taylor and Francis.

Newman, J., M. Brindley, and K. Turner. 2016. *Critical Issues Paper – Dairy Industry in Tasmania*. Launceston, TAS: Regional Development Australia (Tasmania).

Pequeno, R., and D. Elias. 2015. (Re)estruturação Urbana e Desigualdades Socioespaciais em Região e Cidade do Agronegócio. *GEOgraphia* 17, (35): 10–35.

Pink, K. 2003. *Winds of Change: A History of Woolnorth*. Timaru, NZ: Van Dieman's Land Company.

Pink, K., and A. Ebdon. 1988. *Beyond the Ramparts: A Bicentennial History of Circular Head, Tasmania*. Hobart, TAS: Mercury-Walch.

Pritchard, B., D. Burch, and G. Lawrence. 2007. Neither 'Family' nor 'Corporate' Farming: Australian Tomato Growers as Family Farm Entrepreneurs. *Journal of Rural Studies* 23, (1): 75–87.

Robinson, G. M., and D. A. Carson, eds. 2015. *Handbook on the Globalisation of Agriculture.* Cheltenham, UK: Edward Elgar.

Sharp, A. 2010. The Future Comes to Town. *The Age*, October 7, 2010. EBSCO Newspaper Source Plus.

Tasmanian Government. 2018. *Circular Head Regional Economic Development Working Group: Final Report.* Hobart: Tasmanian Government.

The Mercury. 2016. VDL Cream of Crop. *The Mercury*, October 27, 2016. EBSCO Newspaper Source Plus.

TRASE. 2020. Transparent Supply Chains for Sustainable Economies. Accessed July 23, 2020. https://trase.earth/

Trojahn, M. 2020. A Face do Desemprego em Dom Pedrito. *Qwerty*, March 5, 2020. https://www.qwerty.com.br/2020/03/05/a-face-do-desemprego-em-dom-pedrito/

Woods, M. 2007. Engaging the Global Countryside: Globalization, Hybridity and the Reconstitution of Rural Place. *Progress in Human Geography* 31, (4): 485–507.

Woods, M. 2014. Family Farming in the Global Countryside. *Anthropological Notebooks* 20, (3): 31–48.

3 A Change in the role of women in the rural area of Southeastern Anatolia, Turkey

Suk-Kyeong Kang

Introduction

Turkey is estimated to be the seventh largest producer of agricultural products on the global markets and is a country that exports significant quantities of nuts, dried fruits, and fresh vegetables to the European Union, Iraq, the Russian Federation, and the United States. Agricultural products account for more than 10% of Turkey's total exports, and 19% of the workforce is employed in the agricultural sector (FAO 2016, 1; OECD 2019, Chapter 25). However, half of the agricultural workers are estimated to be seasonal agricultural workers, and over a million migrants move for seasonal work with their families every year within Turkey (Çelik et al. 2015, 7; Öz and Bulut 2013, 95).

In Turkey, the reason for such a large increase in the number of seasonal agricultural workers is related to the mechanization of agriculture and the conversion of subsistence crops into cash crops in rural communities where the distribution of land ownership is particularly unbalanced. In the 1950s, mechanization of agriculture in rural areas reduced labor demand, and those who did not own land, or owned small areas of land not suitable for agriculture, became paid agricultural workers or migrated to large cities for paid employment (Kıray 1999; Öz and Bulut 2013, 95). Since then, neo-liberal national policies in the 1980s weakened subsidies for the protection of small-scale agricultural producers, and as the price of inputs for agricultural production increased, many small-scale farmers abandoned agriculture (Aydın 2002, 191). In addition, small and medium-sized farmers began to shift from low-value field crops, such as wheat, barley, or lentils to cash crops with high profitability, such as cotton, nuts, fruit, and vegetables. This change has increased demand for agricultural workers at certain times of the year, such as during weeding and harvesting (mainly from April to September). This requires intensive labor, and as a result, migration of seasonal agricultural labor has increased (Pelek 2019, 607–8). In general, seasonal agricultural workers have adopted an economic pattern in which they move from their hometowns to other areas and work there with their families during the busy farming seasons and then return to their hometowns during the off-farming season where they expend the income that they have earned.

Seasonal agricultural workers are geographically concentrated in the Eastern, Southeastern, and Central Anatolia Regions of Turkey. Among these regions,

DOI: 10.4324/9781003110095-4

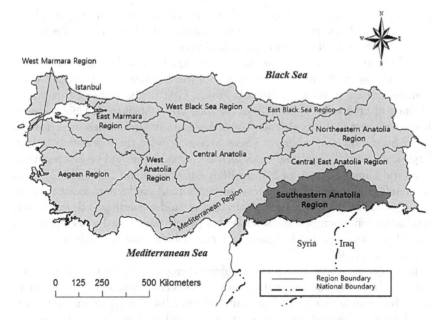

Figure 3.1 NUTS* statistical regions of Turkey.

Source: Author (ARCGIS 10.2).

Southeastern Anatolia accounts for 90% of Turkey's seasonal agricultural work-ers, and many of the local residents make their living from seasonal agricultural work without having other jobs (Figure 3.1).

In terms of land ownership, Southeastern Anatolia is the region with the highest proportion of landless people in Turkey, with a peak in Şanlıurfa Prov-ince (Çelik et al. 2015, 8–10; Oral 2006, 48; Öz and Bulut 2013, 95). To examine why this situation exists in this region, and why residents of the region are forced to make a living as seasonal agricultural workers, it is initially necessary to under-stand the region's geographical environment and traditional social structure.

The Southeastern Anatolia Region borders both Syria and Iraq, and its semi-arid climate is not favorable for settlement or agricultural production. Therefore, livelihoods in this region traditionally favored nomadic or semi-nomadic live-stock herding, including sheep and goats, and subsistence farming where crops such as wheat and barley were grown on dry land. In addition, and in order to overcome the unfavorable geographic conditions and to cope with external threats, a social structure centered on tribal groups emerged. In this social struc-ture, people with common ancestries, or blood ties, lived as tribal groups (*aşiret*) with loyalty to one leader (*aşiret reisi*). In the case of Şanlıurfa Province, 81% of urban residents and 93% of rural residents belong to a tribe (Gökalp 1992, 46; İçli et al. 2012, 20–1; Kang 2013, 88–93). The status of the tribal group to which people belong greatly influences the social, economic, and political aspects of their lives. For example, large areas of agricultural land are controlled by a small number of tribal landowners and noblemen. In addition to agriculture, many of the factories and commercial enterprises in this region that have been established with the progress of industrialization and urbanization are managed exclusively by a small number of these landowners.

In 1995, the Turkish government constructed a large-scale irrigation canal in the Southeastern Anatolia region as part of the Southeastern Anatolia Development Project (*Güneydoğu Anadolu Projesi*, GAP). The General Directorate of State of Hydraulic Works (*Devlet Su İşleri Genel Müdürlüğü*, DSİ) provides irrigation water to the local Water User's Associations, and farmers use irrigation water by paying usage charges. These systems provide the foundation for the large-scale cultivation of cash crops such as cotton, maize, and vegetables in the semi-arid climatic region of Southeastern Anatolia (Kang 2013, 101–2). This has resulted in a significant increase in agricultural production and income in the region, but the unequal concentration of land ownership and agricultural mechanization amongst a small number of people has created a discrepancy in agricultural incomes that is significant for most local residents (İçli et al. 2012, 21; Kang 2013, 88).

Many people whose land is not suitable for crop farming or who do not own any land have become seasonal agricultural workers. These groups often travel with their families to Turkey's agricultural-specialist regions to undertake seasonal work. In general, when seasonal agricultural workers seek work, they use brokers who are individually connected to the large farm owners. These brokers receive brokerage commissions from the seasonal workers and manage their employment at the farms that they represent. Contracts with these farms are not continuous but depend on the agricultural conditions of the farms each year. Therefore, seasonal agricultural workers seek out temporary employment by moving in search of work each year. Generally, those who practice seasonal agricultural work have low status within the social hierarchy of the region (Dedeoğlu 2018, 40; Dedeoğlu and Bayraktar 2020, 176).

Moreover, the recent influx of Syrian refugees to Turkey has had a serious impact on the lives of Turkish seasonal workers. A large influx of Syrian refugees has resulted in an increase in labor supply in the seasonal agricultural market and brought about a decrease in wages for Turkish seasonal workers. In addition, since all these groups reside temporarily in the same locations during the period of seasonal work, friction and conflicts between the groups occur frequently (Pelek 2019, 621–4). The perception of seasonal agricultural workers is gradually worsening as local residents increasingly conflate the two groups, often treating all seasonal workers as refugees, rather than as fellow Turkish citizens. Turkish workers, therefore, report feeling an increasing sense of psychological alienation and deprivation.

These changes in the rural environment have had a particularly significant impact on the lives of women. Female workers are responsible for household activities as well as agricultural labor. During these months they are exposed to unhygienic conditions, such as unclean water, and temporary bathrooms and toilets. Although many socially vulnerable rural women generate cash income and contribute to family economic activities through their agricultural labor, the existing patriarchal social culture makes it difficult for them to increase their economic independence and improve their quality of life.

Female seasonal agricultural workers in Southeastern Turkey

In adopting the Swiss Civil Code in 1926, the Turkish Civil Code attempted to eliminate discrimination and discriminatory legal treatment for women and to change women's social, economic, and political status. However, despite a long period of effort, there are still a number of issues perpetuating to gender inequality and women's low social status in Turkey (World Bank 1993, 1; Yildirim 2005. 350). In particular, the southeastern region of Turkey is still heavily influenced by conservative social and cultural traditions and has a semi-feudal agricultural structure (Bilgen 2019, 536; Ulutaş 2019, 275). Furthermore, the regions' nomadic or semi-nomadic traditions include a strong kinship system (Aşiret), a patrilineal descent tradition, and patriarchal norms. Patriarchy refers to an institutionalized system by which male dominance is manifested in the family and society (Lerner 1986, 239; Solati 2017, 4). In a patriarchal society, women have low status, and the wife's primary obligations are to maintain the family, raise children, and obey her husband (Solati 2017, 55). Women living in this region are strongly influenced by these patriarchal, social, and cultural traditions. Although polygamy is prohibited under Turkish civil law (Turkish Criminal Code 2004; Yildirim 2005, 357; UNHCR 2020), it (*kumalık*) is still practiced in this region, especially in rural areas. Women engaged in seasonal agricultural labor are primarily from low socioeconomic status groups. They migrate with their families and engage in agricultural work while also providing childcare and undertaking housework in inadequate temporary living environments. Under the traditional patriarchal system, wages are paid not to individuals but to the head of the household. Since most women are not directly paid wages if they engage in agricultural work, it is very difficult for women to achieve economic independence (Çelik et al. 2015, 7–31; Çınar 2014, 181; World Bank 1993, 50).

Within Turkey, the poor working and living conditions of seasonal agricultural workers are recognized as a serious social problem. Since 2010, the Turkish government has carried out "The Project for the Improvement of the Working and Living Conditions Lives of Seasonal Migratory Agricultural Workers (*Mevsimlik Gezici Tarım İşçilerinin Çalışma ve Sosyal Hayatlarının iyileştirilmesi projesi, METİP*)". This project seeks to improve workers' health, education, transportation, shelter, labor, and social security conditions, etc. However, this project is not operational in all parts of Turkey. Seasonal workers move to several provinces in search of work, including farms in the provinces where this project has not yet been implemented. In several locations, the project has built public toilets, kitchens, and shower facilities for seasonal workers. However, the workers migrate over the course of the season. As a result, many facilities become neglected and, in several of the temporary residential areas, facilities are no longer available, and workers must resort to temporary/tent toilets and bathing facilities. Adequate management of such facilities is a key challenge to the success of these projects.

In order to not only improve the physical environment but also to improve the quality of life of female workers, it is necessary to examine the daily lives of women working in these areas and to consider how they are affected by the traditional social organization and customs of the region.

Methodological considerations

As Turkey's agricultural base shifts from subsistence to commercial agriculture, and as mechanization increases, many rural residents of the Southeastern Anatolia region, who previously made a living as traditional peasant farmers, are now working as seasonal agricultural laborers. These changes in the rural economic environment have had a significant influence on the lives of women in this region. During the farming season, women who had been mainly engaged in household chores become actively engaged in agricultural work. This study examines the changing social, cultural, and physical environments of rural women who have experienced this shift from household domestic labor to seasonal agricultural work. The study utilizes semi-structured, face-to-face in-depth interviews with female seasonal agricultural workers which were conducted from September 2018 to June 2019. The age of the interviewees ranged from 12 to 60, and the interviews took about 30 minutes per person.

Two specific localities comprise the study area, Şanlıurfa Province and Çukurkuyu town in Niğde Province. Şanlıurfa Province has the highest percentage of seasonal workers in Turkey. In this study, we interviewed households engaged in seasonal agricultural labor, seeking to understand their reasons for choosing seasonal agricultural labor as a livelihood strategy, their backgrounds, and their geographic routes for seasonal agricultural work. In the field survey, in-depth interviews were conducted in September 2018 with 27 female workers and seven male household heads who were making a living through seasonal agricultural labor. The interviews examined geographical information on seasonal work types and geographical migration routes traveled for the two to four months of seasonal agricultural work during the previous year (2017). In addition, the interviews investigated the physical environments of the dwelling units occupied by families during the period of non-seasonal labor during the year and obtained basic data on the problems of these temporary residences.

The second survey area selected was Çukurkuyu town in Niğde Province. According to Turkish Institute of Statistics data (TÜİK 2019), Niğde Province produced 16.01% (732,000 tons) of the country's total potato production in 2019, and in case of white cabbage produced 20.9% (118,593 tons) of the total production. Niğde Province is a specialized agricultural region producing the largest amount of potatoes and white cabbage in Turkey. In addition, the region produced 12.1% (438,327 tons) of the country's total production of apples. This is the third largest amount among the apple-producing provinces in Turkey. Çukurkuyu town is a rural area with a total population of 2,140 as of 2019. Although the number of households in this town is approximately 1,000, only 600 households are permanent residents. The remaining 400 households only reside here for a short period in the summer (Deveci and Gönen 2017, 3). With the introduction of irrigation facilities in 1991, this town began cultivating large quantities of fruit and vegetables such as apples, tomatoes, and watermelons. However, the lack of agricultural labor compared to the high productivity of crops has become a serious social problem in the town. Farmers in the town had to hire seasonal agricultural workers from outside during the farming season and,

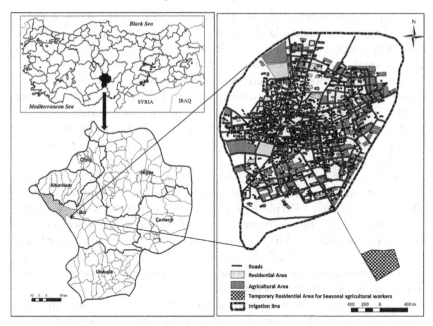

Figure 3.2 Location of study area (Çukurkuyu town, Niğde Province, Turkey).
Source: Author (ArcGIS 10.2).

as a result, they hired short-term migrant workers who could stay in the town and do intensive agricultural work for a short period (2–4 months). In 2016, on the outskirts of the town, a temporary residential area for short-term migrant agricultural workers was created about 1.7 kilometers southeast of the town center, as shown in Figure 3.2.

The new groups of seasonal workers can be roughly divided into two groups. The first group includes workers who migrated short-term with their families for seasonal work from other parts of Turkey, and the second group is comprised of Syrian refugee workers. Of these groups, most of the Turkish seasonal agricultural workers are migrant workers from Şanlıurfa Province, and for that reason, the town was selected as a field study area to investigate female agricultural workers who migrated short-term for seasonal agricultural work from Şanlıurfa Province.

The field study in Çukurkuyu town included in-depth interviews of 57 female workers who temporarily moved from Şalıurfa Province to Çukurkuyu town for seasonal agricultural work in June 2019. Our interviews focused on the types of seasonal agricultural labor that they undertook during the previous year (2018) and the migration routes selected for their work. In addition, field surveys were conducted to assess their living conditions during the seasonal agricultural work period and the social, cultural, and economic difficulties that they faced in their temporary residences. The influx of Syrian refugees into the seasonal agricultural labor market of Turkey provided a further dimension of change, which created new challenges, and this was also investigated.

This study examines how the social, cultural, and economic status of women changed as the economic role of women changed in those rural areas of Turkey where a patriarchal tradition was still dominant. In addition, the problems that arose as female workers undertook agricultural work, and the resulting physical, familial, and psychological burdens were examined. Since the women were often unable to respond freely and honestly to open-ended questions due to the characteristics of the patriarchal community, we induced more honest answers by inserting specific questions so that we could diagnose the real situation of women in a separate space with only women. The following four issues were utilized as research indicators.

First, this study examined the geographical migration routes followed during the busy farming season (mainly from April to September). The repeated moves undertaken over this period placed a great burden on women's lives. In addition, this study compared the temporary living conditions in which seasonal workers lived for migrant labor and the residential conditions in their permanent homes in Şanlıurfa Province. Through this comparison, we investigated the kinds of living environments that they are exposed to every year as domestic labor sources and as seasonal agricultural workers.

Second, we probed the traditional culture of this local community, including basic information on the existence of polygamy, the educational levels of women, and the number of children in the household. As a further indicator of women's status, we asked whether ready communication with the outside world was taking place. To do so, we examined whether female workers possessed mobile phones.

Third, we examined the extent to which the patriarchal culture existed in seasonal agricultural working families. As mentioned above, within patriarchal culture, women are required to obey their husbands and to be dedicated to child-rearing and housework. Today, most women working in the region, who work for wages outside of the home are engaged in intense agricultural work with men who are their family members. Patriarchal culture affects the share of household chores undertaken by men and women in the home. Since no respondents reported that male household members participated in household work (cooking, laundry, cleaning, childcare, etc.), we focused on the Turkish tea culture, which is one of the most frequent and repetitive household chores in which men are likely to participate.

The tea culture is an iconic feature of Turkish culture; however, it is one in which women traditionally prepare and serve tea to male household members, constraining the time available to them for other household tasks and for rest. With the dual responsibility of household and childcare, combined with agricultural labor, the serving of tea becomes symbolic of the challenges faced by women in these households. If men merely sit and wait for women to provide their tea, it will be quite difficult for them to expect men to do housework in a family where this patriarchal tradition is strong. Therefore, this study focused on the tea culture, specifically on whether men brewed their own tea at home. This was taken as an indicator of men's likelihood of sharing household chores. This is based on the basic hypothesis that it would be very unlikely that male household members who do not make their own tea would do other household chores.

Finally, in order to investigate the economic independence of female work-ers, we investigated the amount of cash women had at the time of the inter-view, whether female workers directly received labor wages, and whether they were buying sanitary products. We used the purchase of sanitary products as an indicator of the freedom of women to spend household money on their own specific needs. Sanitary products are essential consumables for women, would be purchased solely by women, and are therefore an indicator of the level of financial agency within the household. In the following sections, we discuss how each of these four dimensions contributes to our understanding of the changing role of women in migratory agricultural labor households in Turkey.

Geographical mobility routes for the livelihood of seasonal workers

Seasonal agricultural workers migrate with their families to places where there is a demand for labor during the busy farming season. The farms where they obtain seasonal agricultural work vary from year to year. Therefore, when a family finds a farm with labor needs, they move to that region. In general, seasonal workers do not work on a single farm during the seasonal work period. Because different crop types have different harvest times, when the family finishes their agricul-tural work on one farm, they move to another farm.

As can be seen in Figure 3.3 and Table 3.1, families work seasonally for about three months, moving between places as different crops are harvested. The families travel up to 1,000 kilometers from their hometowns, usually by rented minibuses.

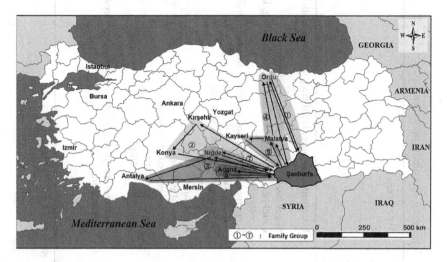

Figure 3.3 Respondents' seasonal migration routes.

Source: Author's fieldwork, September 2018.

Table 3.1 Seasonal agricultural activities conducted by the respondents in the summer of 2017

Family	Type of work	Province	Period	Number of family members		Number of working-age (15–64) family members	
				Female	Male	Female	Male
1	Hazelnuts: pruning, fertilizing, spraying pesticide, weeding, etc.	Ordu	April–July (4 months)	3	3	2	2
2	Sugar beet: hoeing, weeding, watering, etc. Chickpeas, white beans, lentils, etc.: sowing, hoeing, weeding, watering, etc.	Kırşehir Konya	April–August (5 months)	3	5	3	3
3	Tomatoes, peppers and potatoes: hoeing, weeding, watering, and harvesting	Niğde	April–July (4 months)	5	7	5	5
4	Greenhouse cultivation work Apricot picking Hazelnut picking	Antalya Malatya Ordu	July–August (2 months)	3	3	3	3
5	Apricot picking Sunflower seed harvesting Apple harvesting	Malatya Kayseri	July–September (3 months)	4	3	4	2
6	Onion harvesting	Adana	June–August (3 months)	4	3	3	1
7	Tomatoes, peppers, and potatoes: hoeing, weeding, watering, and harvesting	Niğde	April–June (3 months)	4	2	3	1
Total				26	26	23	17

Source: Author's fieldwork, September 2018.

An example of a seasonal worker's geographic migration route follows:

> Because I don't have any job, I go to other provinces with my family and do seasonal work in the summer, especially from June to September. In June, we work to harvest apricots in Malatya province, and then when the work is finished, we move to Kayseri province to harvest sunflower seeds and apples. We work in Kayseri province for about two months, and then return to our home in Şanlıurfa province.
>
> (Female worker, aged 24, September 2018)

The respondents undertook seasonal agricultural labor for economic reasons:

> We go to other cities to do seasonal agricultural work because we don't have our own land and any other jobs.
>
> (Male worker, aged 53, September 2018)

> Since we don't work in winter, we spend more than half of the money that we earned through the summer work in winter. And we save the rest of the money.
>
> (Male worker, aged 45, September 2018)

> We have many family members, but nobody has a job. Also, we don't have our own land for farming. Therefore, my family faces a very difficult economic situation. We are migrating to other provinces due to the lack of job opportunities and low wages in our own town. We have to find work even in other regions to make a living.
>
> (Female worker, aged 25, September 2018)

Migrant workers characteristically set up tents, for whole families at the edge of towns or on working farms. Most of these temporary residences do not have toilets and baths, electricity, or water (Figure 3.4).

In Turkey, bread is a staple food. Given that seasonal workers live outdoors, they mainly make and eat *Yufka* (Turkish flatbread), which can be dried and preserved for a long time and can be simply eaten by wrapping with vegetables or meat. However, a female worker who is undertaking agricultural work on a farm needs time to prepare bread for the family unit. It is not easy to make and prepare enough *Yufka* or bread for an average seven- to eight-member family to consume daily. In fact, female workers in the study area responded that the hardest part of housework is making *Yufka* or bread, followed by laundry (Figure 3.4f).

Figure 3.5 shows the permanent home of a seasonal agricultural worker. As shown in Figures 3.4 and 3.5, while living conditions in the farming season are basic, during the off-season, these workers live like a typical Turkish family

Figure 3.4 Temporary living conditions for families engaged in seasonal agricultural work.

Source: Author, Kırşehir (September 2018) and Niğde Province (June 2019) ((A) temporary tent, (B) tent interior, (C) water tank, (D) temporary toilet, (E) temporary bath, (F) a woman making Turkish flatbread (*Yufka*)). Source: Author.

Figure 3.5 Living conditions of the families of seasonal agricultural workers in their hometowns.

Source: Author, şanlıurfa Province, February. 2018. ((A) dwelling house in a hometown, (B) kitchen, (C) bedroom, (D) toilet, (E) bathroom, (F) living room). Source: Author.

(Figure 3.5). In this type of home, a female worker does not have much difficulty with domestic work:

> When we don't do seasonal work, we live in our home like normal people. By the way, it seems that many people think we are like refugees wandering without a home. After Syrian refugees came to Turkey, it seems that the attitude of the local residents living in the places where we migrated for seasonal work toward us has worsened. These days, Syrian refugees also do a lot of agricultural work like us. Because they work at low wages, our wages have also been lowered. Sometimes they set up tents in temporary living areas and live by sharing the areas with us. But, they are so loud every night that it's hard to live with them together.
>
> (Female worker, aged 25, September 2018)

As such, female workers who take on seasonal farming employment experience significant physical and psychological burdens resulting from the repeated changes in their place of residence every year, from the poor conditions in their temporary residences, and from the attitudes of people who treat them like refugees. In addition, having to live in shared camping areas with Syrian refugees from different cultures is a further recent stress factor.

Female workers in a closed traditional society

In Turkey, the requirement that primary school is compulsory for every citizen has also been included in all constitutions since that of 1876 (Başaran 1996; Özaydınlık 2014, 97).

Article 42 of the Constitution of the Republic of Turkey states that "No one shall be deprived of the right of education. Primary education is compulsory for all citizens of both sexes and is free of charge in state schools" (Grand National Assembly of Turkey, 2010). However, girls are a disadvantaged group in terms of exercising their right to education (Özaydınlık 2014, 97). In particular, Southeastern Anatolia, where Şanlıurfa Province is located, is a region with a low percentage of girls receiving primary and secondary education (Kocabaş et al. 2004, 6–13). In our study area, the education level of female workers was also low. Of the 55 respondents, 31 (57%) female workers were illiterate and with limited education (Figure 3.6).

In terms of polygamy, five out of 22 married women responded that their husbands had two wives, and six of the respondents said that their husbands had three wives. Female respondents expressed discomfort in referring to their polygamous situations and reported that they do not do seasonal agricultural work with the other wives of their husbands and that the wives all reside in different homes in their hometowns. The ongoing presence of polygamy in the study area and among our respondents negatively affects the psychological health of polygamous wives who experience higher rates of psychiatric conditions such as post-traumatic stress disorder and panic disorder (Yılmaz and Tamam 2018, 824).

Şanlıurfa Province is the region with the highest birth rate in Turkey. According to a survey by the Turkish Statistical Institute (TÜİK 2019), Turkey's total

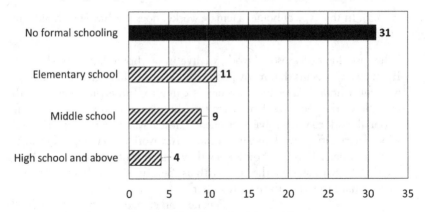

Figure 3.6 The educational levels of female workers (*N* = 55).

Source: Author's fieldwork (Çukurkuyu town, Niğde Province, June 2019).

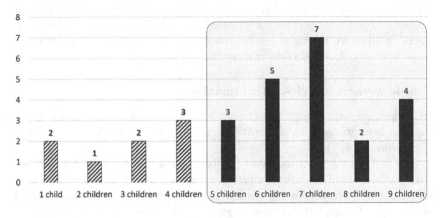

Figure 3.7 Number of children in respondent families (*N* = 29).

Source: Author's fieldwork (Çukurkuyu town, Niğde Province, June 2019).

fertility rate (TFR) was 1.8 in 2019. By comparison, the TFR of Sanliurfa is 3.89. Twenty-one of the 29 respondents with children (72%) in our survey had five or more children of which four each had nine children (Figure 3.7).

Despite having so many children, female seasonal agricultural workers are in a situation where they also do childcare and housework at the same time as agricultural work. If the family members have young children, these are left with the elderly or other family members who are unable to do agricultural work, remain in the temporary residence tents while the remainder go to work on the farm:

> When I was 14 years old, I got married. I have six children now, and I am currently 3 months pregnant, but I am working in the field. The agricultural work starts at 7 am and ends at 5 pm. While I'm working, my thirteen-year-old son, who is weak and unable to do agricultural work is taking care of

three of my younger kids. My husband didn't come this time, and he's doing other things to make money in our hometown. Only I and my children came here with other relatives for doing agricultural work.

(Female worker, aged 34, June 2019)

Despite being pregnant, female workers were working in agriculture all day, and many female workers left their young children with the elderly or children under the age of 15 for care. When the 53 female workers were asked about the age at which they started farm work, 26 (49%) started working before they were 15 years old and eight before they were 10 years old.

Turkey is a member of the Council of Europe (CoE), the International Labor Organization (ILO), and its child labor laws comply with international labor standards (Davran et al. 2014, 1188). According to Article 71 of the Labor Law of Turkey, it is forbidden to employ children under the age of 15 years. However, for the children who are 14 years of age and who have completed compulsory primary education, light work can be permitted as long as the work is not adversely affecting their development, health, or education (Çifçi et al. 2017, 29).

In the case of school-age children of seasonal workers, their school education is affected when the entire family moves to other regions during the semester. Therefore, the Turkish government has established a system that allows these children to easily enroll in local schools if they temporarily migrate to other regions. In addition, cash support, stationery, clothing and food, and fuel are provided to families with school-age children. However, despite this support policy, many school-age children of seasonal workers do not attend school (Davran et al. 2014, 1188).

According to the field survey, because the mothers and all the adults who can work go to the farms to work, it was often very difficult for younger students in the lower grades of elementary school to go to school alone. In some families, older school-age children were left to care for their younger siblings while their mother worked on the farm. In a situation where children are caring for children, the older siblings are less likely to attend local schools. Respondents stated that, even if the children went to school from their temporary residences, they felt a sense of relative distance from the existing students at the school and had difficulty in adjusting to school life. Because of this, many seasonal workers' children did not want to go to school on their own but rather tended to stay in the temporary residences and hang out with children of their own age. Such a residential environment creates a situation in which many children under the age of 15 do not go to school and naturally follow their parents to seasonal farm work. In some regions, there are temporary schools created to directly provide educational opportunities to the children in temporary residences. However, as mentioned above, the METİP project is not currently implemented in all regions.

Female workers in the study area worked on an average of nine hours of agricultural work from 7 am to 5 pm, excluding a 1 hour of lunch break. They were moved by minibus because they often worked on farms far from the temporary residences. Furthermore, 23 out of 54 female workers (43%) did not possess a mobile phone to communicate with in the event of urgent work or emergency situations related to their outdoor work or to their children.

Patriarchal culture examined through tea culture

It is generally assumed that patriarchy is strong in the Middle East and North Africa. This region has a strong kinship system and patriarchal norms. However, there is no quantifiable measure by which to assess this assumption of patriarchy (Solati 2017, 5–17). The rural areas of Southeastern Anatolia Turkey are characterized by traditional sociocultural structures and patriarchal norms (Bilgen 2019, 536). In this patriarchal system, the division of roles between men and women is rigidly divided, and women are required to take full responsibility for household chores and to obey their husbands (Solati 2017, 55).

In order to examine how the patriarchal family environment has changed as the role of women in and outside the home has changed, the survey examined how household chores were shared in the female worker's homes. In order to examine this in detail, we used how the tea service was performed within families as an indicator.

Therefore, taking into account that it is an important element of Turkish culture to drink black tea after a day's work, we investigated whether men prepared tea for themselves after agricultural labor. This was used as an index to examine the degree of patriarchal family culture in working families. As a result of the survey, 74% of households of 54 female respondents said that men do not prepare their own tea. In addition, 69% of respondents said that simply preparing tea was not enough, and that tea should be repeatedly poured into the men's cups by the women until the men do not want to continue to drink tea.

> My husband not only doesn't brew tea, but he doesn't even pour tea from a teapot into his own teacup. Whenever his teacup is empty, I have to pour tea into his teacup for him.
>
> (Female worker, aged 35, June 2019)

In general, Turkish tea culture does not end with a cup of tea. Having tea includes talking while continuing to pour black tea into a small cup in one seat. Therefore, when a person's teacup is emptied, the empty cup has to be refilled until the person does not want to drink any more. As can be seen in the tea culture, female workers have to devote themselves to household chores even after agricultural work, and the patriarchal traditional social culture still has a great influence on their lives. As such, in the patriarchal tradition, even if men and women do the same agricultural work, women remain in sole charge of child-rearing and housework.

Agricultural working conditions and the economic independence of female workers

Every year, the average daily wages of seasonal agricultural workers are compiled based on the wage data from the agricultural holding of the Turkish government (MuhasebeNews 2020). Table 3.2 shows the average daily wages of male and female seasonal agricultural workers between the years 2015 and 2018.

Table 3.2 Average daily wages of seasonal agricultural workers (TRY[a])

Year	Female	Male	Average
2015	46	59	52
2016	53	66	59
2017	60	73	66
2018	67	82	74

Source: Agricultural Holdings Labor Wage Structure (TÜİK 2015–2018).
a Exchange rate: 10 TRY = 1.75 USD (July 11, 2019).

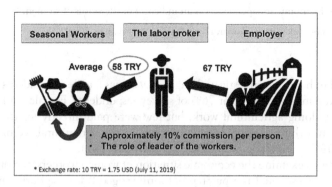

Figure 3.8 Actual daily wages of female workers.

Source: Author's fieldwork (Çukurkuyu town, Niğde Province, June 2019).

Figure 3.8 is a schematic diagram of the survey results based on the field survey in 2019. Labor brokers negotiated between seasonal workers and employers, and these brokers received a 10% commission from the worker or employer in return. As a result, the wages of women workers should be 67 TRY, but the average wage received by them was 58 TRY (Figure 3.8). However, even this amount was not received directly by female workers. According to interviews with female workers, of the 52 female workers, only five women received their own wages, and for 47 female workers (90%) wages were paid to the male householders instead of to the women.

> I work 12 hours a day, and my husband receives his and my wages while I work. If I need money, I get it from my husband.
>
> (Female, aged 44, June 2019)

Of the female workers who received direct labor wages, two were widows and one was a single worker without a husband. As a result, it was confirmed that only two married women were paid directly.

Given that the female workers had been traveling with their families for approximately four months, at the time of the field survey, the amount of money

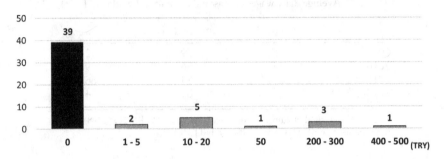

How much money do you have in your pocket right now? (N = 51)

Figure 3.9 Economic independence as an indicator of the status of women.

Source: Author's fieldwork (Çukurkuyu town, Niğde Province, June 2019).

they had in hand was investigated. Figure 3.9 shows female respondents with no money in hand accounted for 76% of survey respondents. Female workers earn money by doing agricultural work, but few were paid directly for their labor. More than a third of unpaid family workers in the agricultural sector of rural Turkey are women (İlkkaracan 2012, 2; Solati 2017, 41).

In order to examine the economic situation of women workers in more detail, this study investigated the purchase of sanitary goods for use during menstruation. This is an essential item for women and we use it in this research as an indicator of economic independence. Out of 48 respondents, 23 (47%) female workers purchased and used sanitary products during menstruation, and 25 (52%) women used old textiles as a substitute for hygiene products. This can cause problems for women's health and hygiene. As such, despite earning money by working on an average of 9 hours of agricultural labor, female workers who were able to buy hygiene products that are personal necessities comprised less than half of the respondents.

Our survey also sought to assess women's desire for economic independence. Thirty-six (69%) out of 52 female workers answered that they would like to receive their own wages. They replied that, if they could receive their own wages, they would like to go shopping, buy clothes, and save for their children. However, of the respondents, 16 (31%) did not want to receive wages directly, they said, "No problem because my husband spending money for the family", "Don't want to keep the money", "I don't need money, because my husband buys for me when I need some goods". As such, some female workers were satisfied with their passive role in household economic activities.

Conclusions

The eastern and western regions of Turkey exhibit wide social, economic, and cultural disparities, and Southeastern Anatolia retains a conservative, traditional patriarchal culture. This study examined the social, cultural, and economic status of female seasonal agricultural workers who are among the most vulnerable inhabitants, both socially and economically, in the region's rural areas. Therefore, the social status of these women does not represent the status of women in most rural areas of Turkey.

The results of our survey become salient in the context of Turkey's recent rural restructuring. The rural areas of this region have been transformed from a self-sufficient and largely subsistence agricultural economy to one based on commercial agriculture as a result of mechanization, neo-liberalism, and the expansion of the irrigation system due to the GAP project. Because of these changes, those rural residents who did not have land to cultivate have moved to work in the cities or in other specialist agricultural areas. Residents moving to the specialist agricultural areas have adopted a short-term migration pattern by which the entire family works in other regions for approximately four months each year and then return to their hometowns.

Our work highlights several key catalysts for this ongoing change. These include difficulties in finding jobs in the rural areas where the respondents live; lack of privately owned land for agricultural activities; low education levels and lack of special skills which prevent these workers from finding other jobs; and finally, large family sizes, including the presence of many children, who become de facto laborers within the family.

As seasonal agricultural work became the main economic activity for these families, many women in the rural areas of this part of Turkey now work not only within the household but also as agricultural workers while traveling with their entire families. However, although many women generate cash income through their agricultural labor and contribute to the family's economic activities, the patriarchal social culture within which they operate has not changed. Due to this patriarchal culture in the family, even if women work outside the home, they often remain in a difficult economic situation and are unable to purchase even essential consumables because male household heads receive all the family's wages. In addition, when women return to their temporary residences after work, they still have to take charge of all the household duties including child-rearing. The lives of these female workers remain quite closed and remain so within a male-centered patriarchal social culture. Their quality of life, as examined by this study's four indicators, therefore remains quite low.

Wider social awareness of the challenges faced by these seasonal agricultural workers remains limited. Although the migrant workers in this study are Turkish citizens and have their own places of residence, they are treated like recent foreign refugees by the wider population because, like the Syrian refugees, they are living in temporary dwellings during the busy farming season. Furthermore, the wages of workers engaged in seasonal farming activities have declined in recent years. Competition for agricultural jobs has increased, as new populations of Syrian refugees compete for the same work. As a result, both conflicts between the national groups and pressure within Turkish seasonal worker households have increased. These social changes have lowered the morale of the seasonal agricultural workers, who are socially underprivileged, and created both economic and psychological distress.

The recent development of information and communication technologies is facilitating changes even in areas where traditional and closed social cultures have hitherto been maintained. Women who were once vulnerable because of their social position are now benefiting from these new technologies and communicating with the outside world. Through this, women's awareness of alternative livelihoods and family situations is gradually growing, and female workers are beginning to complain about their situation.

The poor living and working conditions of seasonal agricultural workers are being increasingly acknowledged as a social problem in Turkey. To address the issues, new projects and research are being conducted to understand and improve the poor conditions. However, the deterioration in the quality of life of seasonal female agricultural workers is not only due to the poor physical environment surrounding them but also to the traditional society, culture, and customs of the region to which they belong. The presence of polygamy, the low education levels of women, high fertility rates, and the patriarchal and closed family culture of this region are quite serious obstacles to guaranteeing the basic rights of female workers. To improve the quality of life of female seasonal agricultural workers, more research and programs focused on women's living and working conditions at the household and community level are needed.

References

Aydın, Z. 2002. The New Right, Structural Adjustment and Turkish Agriculture: Rural Responses and Survival Strategies. *European Journal of Development Research* 14 (2): 183–208.

Başaran, İ. E. 1996. *Eğitim Yönetimi*. Ankara: Yargıcı Matbaası.

Bilgen, A. 2019. The Southeastern Anatolia Project (GAP) in Turkey: An Alternative Perspective on the Major Rationales of GAP. *Journal of Balkan and Near Eastern Studies* 21 (5): 532–52.

Çelik, K., Z. Şimşek, Y. Y. Tar, and A. K. Duman. 2015. *Gezici Mevsimlik Tarım İşinde Çalışan Kadınların Çalışma ve Yaşma Koşullarının irdelenmesi.* 110377, The World Bank, Turkey. Accessed August 3, 2020. http://documents.worldbank.org/curated/en/577571479488978309/pdf/110377-WP-P146215-PUBLIC-TURKISH-kezban-celik-R.pdf

Çifçi, C., E. Bor, and Ö. Özmen. 2017. *Çocuk İşçiliği ile Mücadele Ulusal Programı (2017-2023)*. 63, ÇSGB (*Çalışma ve Sosyal Güvenlik Bakanlığı*). Ankara, Turkey. https://www.ailevecalisma.gov.tr/media/1322/cocukisciligimucadele_2017_2023_tr.pdf

Çınar, S. 2014. *Öteki "Proleterya": De-proletarizasyon ve Mevsimlik Tarım İşçileri.* Ankara: NotaBene Yayınları.

Davran, M. K., M. R. Sevinç, and A. Seçer. 2014. Türkiye'de Mevsimlik Tarım İşçisi Çocuklar (Children of Seasonal Agricultural Workers in Turkey). *XI. Ulusal Tarım Ekonomisi Kongresi*, 1184–92. https://tarekoder.org/2014samsun/58-66-1.pdf

Dedeoğlu, S. 2018. Tarımsal Üretimde Göçmen İşçiler: Yoksulluk Nöbetinden Yoksulların Rekabetine. *Çalışma ve Toplum* 56, 37–67. https://docplayer.biz.tr/104983699-Tarimsal-uretimde-gocmen-isciler-yoksulluk-nobetinden-yoksullarin-rekabetine.html

Dedeoğlu, S., and S. S. Bayraktar. 2020. Bitter Lives on Fertile Lands: Syrian Women's Work and Labor in Turkish Agricultural Production. In *Women, Migration and Asylum* in Turkey: Gender-Sensitivity in Migration Research, Policy and Practice, edited by L. Williams, E. Coşkun, and S. Kaşka, 173–91. Cham: Palgrave Macmillan.

Deveci, E. Ü., and Ç. Gönen. 2017. Application Potential of Gasification Technologies in Rural Settlement Areas. *Eurasian Journal of Environmental Research* 1 (2): 1–7.

FAO (Food and Agriculture Organization of the United Nations). 2016. *National Gender Profile of Agricultural and Rural Livelihoods Turkey: Country Gender Assessment Series*. FAO, Ankara. Accessed August 3, 2020. http://www.fao.org/3/a-i6192e.pdf

Gökalp, Z. 1992. *Kürt Aşiretleri Hakkında Sosyolojik Tetkikler*. Istanbul: Kaynak Yayınları.

Grand National Assembly of Turkey. 2010. *Constitution of the Republic of Turkey*. Article 42. Accessed August 22, 2020. https://global.tbmm.gov.tr/docs/constitution_en.pdf

İçli, T. G., Ş. Ökten, and A. Ö. Boyacıoğlu. 2012. A Study on Hierarchical/Normative Order, Marriage and Family Patterns in Bin Yousuf Tribe of Southeastern Turkey. *Advances in Applied Sociology* 2 (1): 19–29.

İlkkaracan, İ. 2012. Why so Few Women in the Labor Market in Turkey? *Feminist Economics* 18 (1): 1–37.

Kang, S. 2013. Development of Cotton Farming and Transformation of Rural Area in Sanliurfa Prefecture, Turkey. *Journal of the Korean Geographical Society*. 48 (1): 87–111. http://journal.kgeography.or.kr/articles/article/ekKz/

Kıray, M. 1999. Sosyo-Ekonomik Hayatın Değişen Düzeni: Dört Köyün Monografik Karşılaştırılması. In *75 Yılda Köyden şehirlere*, edited by O. Baydar, 151–62. İstanbul: Türkiye Ekonomik ve Toplumsal Tarih Vakfı.

Kocabaş, İ., S. Aladağ, and N. Yavuzalp. 2004. Eğitim sistemimizdeki okullaşma oranlarının analizi. In *XIII. Ulusal Eğitim Bilimleri Kurultayı*, edited by H. Atılgan and I. Çınar, 1–15. Malatya: İnönü Üniversitesi. https://www.pegem.net/dosyalar/dokuman/288.pdf

Lerner, G. 1986. *The Creation of Patriarchy*. New York: Oxford University Press.

MuhasebeNews. 2020. Average Daily Wage of Seasonal Agricultural Workers in Turkey Increased by 17.1% in 2019. *MuhasebeNews*, February 27. Accessed March 5, 2020. https://www.muhasebenews.com/en/average-daily-wage-of-seasonal-agricultural-workers-in-turkey-increased-by-17-1-in-2019/

OECD (Organisation for Economic Cooperation and Development). 2019. *Agricultural Policy Monitoring and Evaluation 2019*. Paris: OECD Publishing. doi:10.1787/39bfe6f3-en.

Oral, N. 2006. *Türkiye Tarımında Kapitalizm ve Sınıflar*. 6, TMMOB Ziraat Mühendisleri Odası. Ankara: Turkey.

Öz, C. S., and E. Bulut. 2013. Mevsimlik Tarım İşçilerinin Türk Hukuk Sistemi İçerisindeki Yeri (The Status of Seasonal Agricultural Workers in Turkish Legislation). *Çalışma Dünyası Dergisi* 1 (1): 94–111.

Özaydınlık, K. 2014. Toplumsal Cinsiyet Temelinde Türkiye'de Kadın ve Eğitim. *Sosyal Politika Çalışmaları Dergisi* 33: 93–112. doi:10.21560/spcd.03093.

Pelek, D. 2019. Syrian Refugees as Seasonal Migrant Workers: Re-Construction of Unequal Power Relations in Turkish Agriculture. *Journal of Refugee Studies* 32 (4): 605–29.

Solati, F. 2017. *Women, Work, and Patriarchy in the Middle East and North Africa*. London: Palgrave Macmillan.

TÜİK (Turkish Statistical Institute). 2019. Bitkisel Üretim İstatistikleri. Accessed August 3. 2020. https://biruni.tuik.gov.tr/medas/?kn=92&locale=tr

Turkish Criminal Code. 2004. *Multiple or Fraudulent Marriage, Religious Marriage Ceremony*. Article 230. Law Nr. 5237. TBMM. (Grand National Assembly of Turkey). Accessed Jun 28, 2018. https://www.tbmm.gov.tr/kanunlar/k5237.html

Ulutaş, E. 2019. Eril kodlar kıskacında kadınlık halleri: Kumalık. In *Aile Kurumuna Farklı Bakışlar*, edited by Ö. A. Sönmez and G. Aksan, 265–99. Konya: Çizgi kitabevi.

World Bank. 1993. *Turkey: Women in Development*. A World Bank Country Study, World Bank, Washington, DC.

Yildirim, S. 2005. Aftermath of a Revolution: A Case Study of Turkish Family Law. *Pace International Law Review* 17: 347–71. http://digitalcommons.pace.edu/pilr/vol17/iss2/8

Yılmaz, E., and L. Tamam. 2018. The Relationship between Polygamy and Psychiatric Disorders in Turkish Women. *International Journal of Social Psychiatry* 64 (8): 821–7.

4 Agricultural Transition in Rural China

Intersections of the Global, National and Local

Guy M Robinson, Bingjie Song, and Zhenshan Xu

Introduction

The development of China from a largely impoverished, underdeveloped country in 1980 to a modern, urban, and industrial powerhouse in 2020 is one of the most dramatic economic transformations of the modern era. In 1980, China had a per capita gross domestic product (GDP) below the average for the developing world, and agriculture accounted for three quarters of the labour force. By 2020, GDP per capita (excluding Hong Kong and Macao) was equivalent to that of Argentina, Mexico, or Turkey (IMF 2020), and agricultural employment was under 15% of the national total. The point at which the proportion of urban residents officially exceeded those living in rural areas was reached in 2008, though the real gross output value of Chinese agriculture has continued to grow at an average rate of 5.4% per annum since 1980 (Huang and Rozelle 2018). This dramatic turnaround has brought huge changes to the Chinese countryside, including massive flows of outmigrants to the fast-expanding cities creating widespread rural depopulation, the growth of urban sprawl swallowing up prime farmland, major environmental deterioration, and agricultural modernisation. Rural transformation has been characterised by the substitution of capital for labour, typified by adoption of modern technology on farms, reforms to marketing and land tenure, and a broadening of the rural economy to embrace privatisation and substantial development of a rural tourism industry.

Many of the features of China's economic 'miracle' are now well established and documented (e.g. Naughton and Tsai 2015). Similarly, some of the key features of the rural transformation have become familiar to many Western observers, including the so-called 'hollowed' villages created by depopulation (Liu et al. 2010; Ye 2018), the 'ghost cities' of the city edge real-estate boom (Shepard 2015), and the rise of 'dragonhead' corporations in the food processing sector (Schneider 2017; Zhang 2019). However, these features are just some of the many different aspects of a complex multifaceted transition from a rural economy that was essentially based on production by a poor collectivised rural peasantry, to an increasingly mechanised, productive modern farming system. It is a transition that has been uneven both spatially and temporally, reflecting the interplay between substantive policy reforms, global processes affecting the newly open Chinese economy and society, and local responses to new opportunities and pressures.

DOI: 10.4324/9781003110095-5

This chapter addresses this transformation of Chinese agriculture, focusing on key elements in the transition to modernisation and the creation of a multifunctional countryside in parts of the country, by examining the interplay between local factors and both national and global forces. It draws upon research undertaken by the authors in the hinterland of a major city, Xi'an, in north-western China and also by other geographers and scholars in allied disciplines to highlight both the uneven nature of the transition and the significant role played by government policy. The latter has provided the dominant framework within which farmers and rural businesses operate, orchestrating the extent to which local and global actors can shape outcomes underpinned by the overriding impress of legislation from central government. However, this legislation has substantially changed direction across the 40-year period that is addressed here, giving rise to the substantive shaping of the transition and the transformation.

Theorising agricultural transition

Written in the late 1890s, Karl Kautsky's *The agrarian question* examined the relationship between the European peasantry, the Great Agricultural Depression, and the rise of both global market integration and international competition (Kautsky 1988). He concluded that the peasantry was not being displaced wholesale by the advance of capitalism, but rather some peasants were intensifying production for the market economy and were engaging with new processing and agri-industrial enterprises that had appropriated some farming functions (Goodman and Redclift 1981). Kautsky's book and related work by Lenin (1982) tended to neglect forces operating external to the agricultural system, though the latter recognised the social differentiation of the peasantry through the impact of the market. He noted the potential of the smallholding peasant economy to become part of the capitalist economic system in a transition from subsistence production to a more differentiated family farming tied not only to evolving local markets but also to provincial and national systems of trade and consumption, with prospects for larger businesses to tap into global networks as China's export trade expands dramatically. Given the character of Chinese agriculture in 1980, overwhelmingly dominated by peasants either farming collectively or on small family-run plots, these theories about the modernisation of this mode of production are highly apposite.

The impact of the market, allied to various forms of government support for family farming, has been cited as the driving force behind agricultural transition in the West, in which there was a gradual move from peasant smallholdings to larger family farms, which after 1945 became firmly productivist in focus (Robinson 2004, 62–64). The term 'productivist' has been applied primarily to signify three structural changes in agriculture: intensification, concentration, and specialisation (Bowler 1986). These changes have been typified by high levels of mechanisation and application of biotechnology, closer links between family farms and both suppliers of inputs and 'downstream' processors and retailers (e.g. via contract production), and the impact of globalisation on the overall agri-food system (Robinson 2018). This has had some negative impacts on the environment (Yu and Wu 2018) and created a dualism between marginalised family farms and large, heavily capitalised agri-businesses, albeit with significant local variation (Ji et al. 2016). Strong government support for agriculture has often typified productivism, especially in the European Union (EU) and the United Kingdom. This is also a key facet of agricultural development that is evident in China where central government has strongly promoted modernisation, with support for mechanisation, grants to support desirable practices (e.g. greater use of artificial fertilisers, improved irrigation systems), and reforms to marketing (Huang and Rozelle 2018).

Farm households in the developed world have had to seek alternative sources of income to survive, with different options apparent, from diversifying farm production to developing a range of new on-farm and off-farm income sources. This broad process of adjustment and restructuring has given rise to the term 'post-productivism', signifying more concern with pro-environmental outcomes from farming (e.g. organic agriculture), a shift of emphasis from quantity to quality in food production, the introduction of alternative farm enterprises, the progressive restructuring of government support for agriculture, and commodification of former agricultural resources (e.g. for tourist development) (Zinda 2017). Some of these elements are also present in Chinese agriculture. However, Wilson (2001, 2007) contends that a more appropriate term for recent changes characterising agriculture in many parts of the world is 'multifunctionality', in which it is recognised that agriculture plays various roles in addition to the production of food and fibre, and so contributes to the development of multifunctional landscapes. Hence, agriculture can protect the natural environment, maintain the rural landscape, and lay the foundation for rural development. Individual farms play a part in this by diversifying into different enterprises, such as farm-based tourism, food processing, and educational visits (Fang and Liu 2015).

Some observers have identified other trends in addition to multifunctionality. These are typified by divergent aspects of individual farm trajectories. At one extreme is a move towards low input–low output farming typified by some hobby and part-time farmers, especially in peri-urban fringes (Zhou et al. 2019). At the other, there is super- or neo-productivism in which the focus on industrial-style farming methods has intensified (Wilson and Burton 2015). In China, the latter has tended to be associated with intensive livestock production, notably in the pig and poultry sectors (Qian et al. 2018). Both industries have been affected by serious outbreaks of disease, which have greatly affected production. The most widespread recent disease has been African swine fever, first affecting Chinese pig herds in August 2018. As many as 55% of the national herd may have died in 2019 (The Pig Site 2020). Many pork producers changed quickly to poultry rearing, but oversupply of chickens in Chinese markets has made the switch an uneconomic proposition in many cases. In the EU and North America, there are more areas where emphasis has been placed on intensifying and increasing production, with new biotechnological advances and computing co-opted by farming in what is increasingly being termed a new agricultural revolution or Agriculture 4.0 (De Clercq et al. 2018).

The policy setting

The context within which Chinese agriculture operates has been largely determined by major central government policy reforms, initially dating from 1978. These have moved the country away from the collectivism that had been dominant since the Chinese Communist Party came to power in 1949, with the creation of the People's Republic. A more free-market economy has been adopted albeit with many distinctive Chinese features, including strong central government controls. Land tenure reforms in the 1980s created the household responsibility system (HRS) in rural areas, whereby land was leased to individual farm households by the collective owners, which were usually a township government or a village committee. Farm families were granted use rights to land, which were subsequently extended for an indefinite period. This means that individual households can make decisions about the land they farm, and they can rent out or sell these use rights. Initially, the village committees allocated the land between households, but arbitrary land exchanges have now largely ceased, though expropriation to enable urban development remains a major issue near growing cities. Most of the plots farmed by a single family are small with much land fragmentation so that over 90% of households farm less than 1 hectare (Tan et al. 2013). In addition, the norm is for a household to farm parcels of land located in different places rather like the situation under excessive 'parcellation' under the Napoleonic Code in Europe. In China, this reflects the land redistribution that generally accompanied the introduction of the HRS, whereby families were allocated different types of land to farm within a village (Lu et al., 2018). This creates inefficiencies and keeps labour productivity low.

To offset some of the limitations posed by the plethora of small farms, government has promoted the growth of cooperatives and the creation of agri-businesses, known as 'dragonheads', which are large agricultural industrialised corporations. Cooperatives have been supported since 1985 and have a bewildering variety of different forms and purposes. Initially, community cooperatives were 'multipurpose organisations responsible for handling the administrative and social affairs of the village; providing agricultural as well as public services; leasing land and co-ordinating land and water use; and organising initiatives to develop the village economy' (Clegg 2006, 223). Supply and marketing cooperatives and rural credit cooperatives also appeared in the 1980s, with attempts to increase their independence and turn them into farmers' cooperatives. They have been followed more recently by various other forms of cooperatives attempting to benefit from scale economies, though many have morphed into private agribusinesses. This has occurred primarily through control of cooperatives being vested in relatively few hands and often involving key local officials who have seen the opportunity to profit from operating as a commercial business rather than a cooperative or by entering into partnership with dragonhead companies.

With the rapid growth of major cities from the 1990s, especially in eastern China, millions of people migrated from the countryside, often leaving behind the elderly to continue farming and to look after children. By 2019, there were 291 million migrants from rural areas living in cities (CLB 2020), and many villages in more remote locations, especially across western China, had been 'hollowed' out, contributing to reduced agricultural productivity. This has been in stark contrast to rural areas in easy reach of the growing cities where the market economy has stimulated production to meet rising consumer demand. Here, the traditional dominance of grain production has often been replaced by diversified farming, with cash crops, horticulture, and livestock more prominent. Farmland in the peri-urban fringe has often been swallowed up by the expanding cities, but advances in productivity through adoption of greater mechanisation and use of purchased inputs has meant that losses of farmland have been partially offset in terms of the maintenance of overall food production. Nevertheless, long-term pressure on the land has brought major environmental problems necessitating major government programmes to address land degradation and deforestation. The largest such scheme, Grain for Green, introduced in 1998, has targeted over 100 million hectares, converting low-yield farmland to forest and afforesting barren mountains and non-productive land, especially on sloping land (defined as land with slopes in excess of 25°), largely in the upper and middle reaches of the Yellow and Yangtze River basins (Chen et al. 2015).

Multifunctional agriculture (MFA) has featured in official policy since 2007 when the 'multiple functions' of agriculture were first posited as playing a major role in rural development. MFA was viewed as one of the means to modernise agriculture and assist in reducing the imbalance between the dynamic, urbanised industrial sector and a lagging rural sector. In 2018, a new Rural Revitalisation Strategy explicitly stated that China should expand the multiple functions of its agricultural sector (Liu 2018). MFA is now considered as conducive to helping solve various rural problems, including maintaining food sovereignty, alleviating poverty, and obtaining more balanced development between the cities and the countryside (Zhao et al. 2019).

The transition to multifunctional agriculture? The example of Xi'an's rural hinterland

Fieldwork was conducted in 2017/18 in the peri-urban fringes of a major city in north-west China, focusing on the changes occurring in the area in the four decades since the initiation of major economic reforms in 1978. Specifically, the study area is White Deer Plain, located c30 to 40 kilometres east of Xi'an, capital of Shaanxi Province (Figure 4.1), with 15 million population living in the urban metropolis. White Deer Plain is in Baqiao District (population 123,000 of whom half live in farm households). Traditionally, this was an area of grain production, but since the early 1990s there has been a wholesale conversion to horticultural production and allied tourism development, including on-farm provision of restaurants, cafes, retailing, and pick-your-own (PYO) schemes. These dramatic changes to land use reflect not only the interplay of various policy drivers but also the changing context in which farming operates, which includes the impress of the urban market, adoption of new technology, and the impact of the growing urban middle class seeking to engage in rural leisure activities. Fieldwork was conducted by two of the authors in late autumn and winter of 2017/18. Six

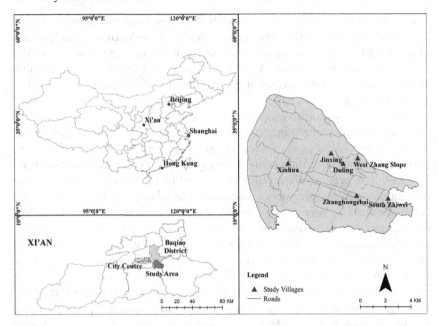

Figure 4.1 The study area: White Deer Plain, Shaanxi Province, North-west China.

Source: Authors.

Table 4.1 The agricultural profile of sample villages in White Deer Plain

Villages	Households (n)	Gross area (ha)	Orchard area (ha)	Rate of renewal of fruit trees (%)	Land transfer area (ha)	Year fruit planting started	Fruit sales (million/ US$)
Duling	240	80	57.3	40	69.3	1998	1.4
Jinxing	290	73.3	66.7	20	26.7	2007	0.9
West Zhang Slope	148	106.7	100	70	—	1987	4.4
South Zhiwei	610	200	46.7	—	13.3	2005	—
Zhanghongzhai	320	93.3	73.3	20	—	2008	1.8
Xinhua	396	197	13.3	—	107.7	NA	—

Source: Authors.

villages across White Deer Plain were selected at random, and village heads and secretaries were briefed on this study (see Figure 4.1 and Table 4.1). In total, 105 semi-structured interviews were conducted with farmers from the selected villages, focusing on household structure, crops cultivated, farmhouse-based tourism, the relationship between local tourism and agriculture, the history of land use policies, tourist development, extent of property renovations, cost of village infrastructure, and agricultural subsidies. In addition, the village heads from three villages and the village secretaries from two villages were also interviewed.

Modernising agriculture

In terms of the agricultural transition, perhaps the most important initial influence was the HRS, first introduced in some rural areas in 1979 and then officially established in 1982, as this gave individual farm households the opportunity to make decisions about what happened on their own small plots of land rather than being part of a collectivised farming system. Their ability to rent use rights on the land of neighbouring farm families provided opportunities for some development of larger holdings, though these were still small by developed world standards. The extensive survey of farm sizes worldwide c2010 by Lowder et al. (2016) referred to farms of <2 hectares as 'small'. These accounted for about 12% of the world's agricultural land. Chinese farms comprised 35% of the world's total farm holdings but only 11% of the agricultural land, and the amount of farmland in China had decreased since 2000, mainly due to urban sprawl. Seventy-two per cent of farms worldwide are smaller than 1 hectare in size. Local government advocated land transfers to realise scale economies and, specifically, to increase the scale of cherry planting in White Deer Plain. Limited by technology and expertise, some farmers had felt constrained by just growing cherries on their own small plots and so began to lease land from other villagers or gained income by renting land to larger enterprises. Some farmers have also taken advantage of the opportunity to combine farming with tourism and so have rented land to expand production, generating more capital to invest in on-farm tourism enterprises:

> In the countryside, each household owns [the use rights] to about half a hectare of land. After the land transfer policy was released, I rented two hectares of land from other villagers with cherries and grapes planted.
>
> (male farmer, aged 51, a leading horticulturalist in Jinxing,
> White Deer Plain)

Rentals were often linked to the migration process in which rural residents of working age increasingly migrated to the cities for work but could rent out some of the use rights to the family's land or leave some land in the hands of their parents to manage. The presence of the One-Child Policy, in place between the late 1970s and 2015, may have contributed to land being rented out by ageing parents as their only children migrated to the cities. The other key factor was the gradual relaxation of state controls over marketing, which gave farmers more scope for selling produce in urban markets. In the case of White Deer Plain, these changes in the 1980s provided opportunities to move away from grain production to help meet rising consumer demand for fresh fruit and vegetables. In the 1990s, there was widespread adoption of the production of cherries and other fruit for the Xi'an market. Farmers and members of farm households would take their harvest and sell it themselves on the streets of the city as well as selling to wholesale markets. Cherries had previously been grown on a small scale in the Plain, but this was now commercial production, which was encouraged by village committees and township governments under the wider auspices of central government entreaties to reduce rural poverty and adopt modern farming practices. The latter have often been supported via grants, acting as incentives for farmers to adopt desired practices.

Commercial cherry production involved some innovative decision-making by the initial pioneering growers, but its rapid spread was sanctioned and promoted by village committees and party cadres who saw it and other horticultural innovations as a means for them to meet government targets regarding poverty alleviation and increased production. However, in terms of the incomes of farm families it was also the rapid growth in the urban middle class, possessing increased disposable income, that provided additional commercial opportunities in the form of provision for tourists. Interviews with cherry growers in one of the villages in White Deer Plain showed how a single innovator led the move to grow cherries commercially in that village:

> In 1990, my home was too poor; we even couldn't afford tuition for my children. At that time, one of my cousins gave me a suggestion. He said maybe planting cherry trees would be a good way to increase income. He is a professor in Northwest A&F University, ...majored in vegetation science. I thought, just try it! What if I could succeed? Finally, I really made it! Cherry trees grow well, and then the villagers were jealous, and followed me to plant cherry trees. But up to now, I still have the most cherry trees!
>
> (male farmer, first commercial cherry grower
> in Duling village, aged 70)

This represented a dramatic and risky change, so the transition away from grain production was only gradual and initially involved seeking advice from experts who confirmed local soils were suitable. This was both an opportunity and a challenge. One of the interviewees described it as, 'a leap into the unknown', especially as it can take from five to eight years before cherry trees bear fruit after they have been planted. Returns may have been increased by the adoption of different varieties, with some support from government grants and slightly different production systems adopted (see Table 4.2). However, change was not necessarily straightforward:

> There is an old idiom in China, 'cherries are delicious, but not easy to be grown.' In 1995, my family planted 50 cherry trees in total, but it was too easy (for the trees) to die. I have to treat them as babies! Even so, I still have changed only 20 of 50 cherry trees until now!
>
> (male farmer, cherry grower, Zhanghongzhai, aged 50)

Table 4.2 Characteristics of cherry production in four sample villages

Villages	Row and column spacing (m)	Numbers (plants/Mu)	Production (kg/Mu)	Costs (US$/Mu)
West Zhang	3 × 4	60	1,000	220
Duling	4 × 6	50	750	670
Jinxing	4 × 6	50	820	650
Zhanghongzhai	4 × 5	40	1,000	580

Source: Authors.
Note. 15 Mu = 1 ha.

Figure 4.2 Grapes being grown by smallholders for JunDu organic grapes in White Deer Plain.

Source: Authors.

Across the Plain, other villages have adopted different horticultural crops, including cherries, wine grapes, strawberries, melons, walnuts, kiwifruit, and vegetables in a horizontally differentiated arrangement geared primarily to supply the local metropolis (Figure 4.2). Although much of the innovation has been part of a farmer-initiated process, there has also been support from government at various levels. This reflects an overarching strategic direction setting by national government as part of an attempt to modernise the farming sector and to link producers to consumers. This has taken various concrete forms, including grants to promote mechanisation, increased use of organic fertilisers, and adoption of higher-yielding varieties. There has also been the establishment of demonstration sites by local government to promote best practices as a means of translating research and development directly to the farmers. In general, the farmers interviewed approved of the support they were receiving and of the positive impacts produced by the new crops. Investment in improved infrastructure in the district has also been beneficial, increasing ease of access of tourists to the area.

> Everyone in the village has benefited from planting cherries and the quality of life has improved. During this process, everyone helps each other and is full of hope for life. The government took into account reasonable requests we put forward, such as the increase in the number of roads to the village from one to three and a ring road.
>
> (Secretary of West Zhang Slope village, male, aged 50)

Farm-gate sales of cherries command a price more than double that of the price when sold to wholesalers. A similar price to that at the farm-gate is obtained from e-sales which have grown in volume in recent years. E-commerce first appeared in the area in 2013 as a new sales model adopted in a single village. It was then prompted by Baqiao District government, which introduced several well-known e-commerce enterprises (e.g. Alibaba, JD.com) and logistics companies (e.g. S.F. Express, EMS, China Post) to the farmers. Subsequently, the scale of e-commerce sales of fruit has increased year by year. This growth of e-commerce is part of substantial changes in the provision of utilities in the study area especially since 2000 (e.g. rural electrification and domestic water supply) and general infrastructure (e.g. tarmacked roads). Improved public transport has increased the ability of residents to work in the nearby city while enjoying higher standards of living and continuing to reside on farms.

In 2016, 670 commercial farmers in the region chose to sell their agricultural products through online platforms so that online sales accounted for 40% of total sales at this time. Among the logistics companies in White Deer Plain, S.F. Express has achieved an annual revenue of 30 million yuan (US$4.4 million). To ensure the freshness and quality of fruit, the company has pioneered cold-chain distribution using refrigerated trucks and incubators. In terms of delivery time, it has established a standard of delivering fruit quickly after harvesting, the following day to first-tier cities (e.g. Beijing, Shanghai) and the same day in Xi'an. Various e-commerce organisations have now infiltrated all the villages in the region, greatly boosting the sales of agricultural products. This contrasts sharply with cherry sales in the past:

> When cherries are picked, it means cherries have to be sold within two days before they rot. In the past, awaiting purchase and the wholesalers were the only way to sell our products and wholesalers always kept the price of our cherries deliberately low. If we didn't compromise with them, then all my cherries would be unsalable.
>
> (female cherry farmer, South Zhiwei, aged 47)

Local government has actively intervened to publicise and promote locally grown cherries to further expand their popularity and influence. For example, the district government has registered 'Baqiao Cherries' as a trademark, taking advantage of the strong development momentum of the cherry industry in White Deer Plain. In addition to cargo partnerships with logistics companies, the government has required producers to switch to standard packaging emblazoned with distinctive logos such as 'Baqiao grapes' and 'Baqiao cherries'. When couriers deliver products, they circulate the agricultural products' leaflet from Baqiao District, which also emphasises the brand and maintains its prominence.

During interviews, farmers commonly said that they would select the best cherries for sale via e-commerce so that more people could readily recognise the high quality of their cherries, and they would be willing to pay premium prices. Farmers who planted and sold high-quality fruit achieved a win-win situation of higher prices and growing reputation. The price transmission mechanism forces farmers to attach importance to quality and its improvement and encourages integration of standardised planting into the whole production process, thus forming a virtuous circle of ecological and financial benefit.

Given the small average size of holding farmed by a household, agriculture was generally combined with other forms of income. This could be income derived from off-farm employment in the city or from local rural industry, as encouraged by government in the 1980s and 1990s, or, increasingly, it has come from other on-farm enterprises. Typically, this has taken the form of income from tourism.

'Farmhouse Joy' tourism

The growth of rural tourism in China has been massive and dramatic in the sense that it has increased very quickly and has attracted substantial numbers of tourists. China's rural areas attracted 2.5 billion tourists in 2018, generating spending of over 1.4 trillion yuan (US$0.4 trillion) according to data from the National Tourism Administration (Zhou 2019). This represents a huge surge in the popularity of rural tourism in the country, as data for 2016 show only just over half of these numbers (Chen et al. 2018). There has been support for the tourism industry from government, both at a national level and through various local programmes. However, it has been the rising wealth of the urban middle class that has driven the surge in numbers, coupled with the availability of more leisure time and changing attitudes towards the countryside.

White Deer Plain has become a noted area for farm-based tourism or *Nong jia le* (meaning 'happy farmer home' or 'farmhouse joy') involving similar elements to those seen for several decades in the West (Park 2014). This tourism is characterised by a growth in farmhouse accommodation, including bed and breakfast, alongside farmhouse cafes and restaurants, farm shops selling farm produce and vernacular products direct to consumers, and provision of on-farm activities, including PYO fruits and vegetables. A key element has been the opportunity to consume 'farmers' foods' (*nongjia cai*) as an 'authentic' rural experience. Urban residents are being offered the opportunity for a brief return to the traditional rural roots of Chinese society as they experience an aspect of farm life directly on or close to a working farm. It is essentially a nostalgic view of rurality that contrasts with the widely held view of many Chinese urbanites that rurality is associated with hardship, deprivation, and lack of choice (Wu 2014; Griffiths and Zeuthen 2014). Across the Plain, it has been PYO cherries that have proved extremely popular as a tourist attraction, especially during the annual May holiday week. Picking cherries and eating a meal on a farm is often part of a day's outing for a family, but there are also various other attractions, both scenic and cultural to entice visitors.

The most visited attraction in the Plain is a 'tourist village' (Bailuyuan – Bailucang scenic area) covering 140 hectares with a total investment of 1 billion yuan (US$158 million) (Robinson and Song 2019) which was opened as part of China's Western Development Strategy (WDS) programme under the 13th five-year plan (2016–2020) of the China National Development and Reform Commission (Figure 4.3). The folk culture experience at the village is based on the famous writer Chen Zhongshi's literary masterpiece 'Bai Luyuan' (White Deer Plain), and hence the name of the scenic area and its locale. This has more than 400 folk buildings, built by professional craftsmen according to traditional construction methods. So, this is a recreated historic village, which has shopping outlets, entertainment, and opportunities for consumption of local produce, thereby acting as a new outlet for nearby farmers. The tourist village is also located along the former Silk Road under China's One Belt One Road initiative launched in 2013 (Huang 2016). It is a good example of several recent major investments in tourism megaprojects, with central government advocating tourism as a key element in rural development strategy and providing a substantial part of the investment

Figure 4.3 Bailuyuan – Bailucang scenic area.

Source: Authors.

(Su 2013). In addition to recreated historic buildings, the attraction offers local folk culture, a hot-air balloon carnival, a music forum, sports experience camp, LightScribe concert, shopping outlets, entertainment, and opportunities to consume local produce. There is also a popular ski run attached to the complex.

> What he [*the author Chen Zhongshi*] describes is our true story, our true life. Many visitors want to see places where the author has lived, worked and created.
>
> (male farmer, aged 53, living in the same village where the author had lived)

Within the broader context of national government support for tourism, provincial and local governments have taken various steps to help grow the tourism industry in White Deer Plain. One of the first measures was the establishment of a cherry festival in 2004, now held annually to mark the time of the cherry harvest. It attracts nearly 1 million visitors each year, adding over 100 million yuan (US$14.6 million) to the local economy. The district government has cooperated with Xi'an municipal government to produce substantial advertising for the festival through television, newspapers, radio, and other channels, encouraging tourists to pick and buy cherries. Meanwhile, to help facilitate *nong jia le*, the Baqiao District government has introduced a door-to-door service of farmers' health licenses and business licenses, simplifying a previously cumbersome process.

As part of measures to attract more tourists, Xi'an municipal government has funded measures to improve the physical appearance of villages across White Deer Plain. These grants are linked to *Guidelines for the Construction of the Beautiful Village* issued by the Ministry of Agriculture and Rural Affairs, which provides quantitative standards for the construction of 'beautiful villages' (MARA 2015). Across the Plain, eight villages have been awarded the rating of 'beautiful', with two (Jinxing and Xiche) designated as model 'beautiful' villages. The *Guidelines* have four main aims: increase green space, produce sealed roads, 'brighten' villages (by redecoration and refurbishment), and beautify the environment.

An important aspect of refurbishment has been a policy termed the *Toilet Revolution* (MCT 2017), which requires farmers operating *nong jia le* to improve the toilets and plumbing arrangements in their households. This involves (1) moving the toilet from the outside of the house to the inside, (2) replacing dry-pail latrines with flush toilets, and (3) dealing with sewage in an environmentally friendly fashion. Each participating household has been compensated by 2,800 yuan (US$408.40) once their plans have been accepted by local officials. It is also the responsibility of the villagers to whitewash their external walls and add greenery around courtyards. At the village level, transformation of the local network of water pipes and improvements to public facilities have been required by the *Guidelines*. This can provide a better basis for supporting tourism enterprises, especially farmhouse-based bed and breakfast (Gu and Ryan 2010; Bao et al. 2015).

Conclusions: agricultural transition in China

White Deer Plain represents a success story in terms of the realisation of central government's broad strategies for agriculture over a 40-year period. The area can be viewed as a microcosm of the myriad changes occurring to China's rural economy and society in the rural–urban fringes of its megacities, especially in the eastern provinces. Elsewhere, and especially further west and south-west from Xi'an, in regions more remote from urban influence, out-migration has been more dominant and the forces of modernisation less pronounced despite major poverty alleviation programmes. Perhaps the most fundamental aspect has been the introduction of the HRS as this has given small farmers a degree of autonomy over the management of the land they operate, albeit with constraints imposed by the small plots and the ongoing role of village committees. We have not explored the latter in depth but it is clear that, across the Plain, the committees and the party cadres have supported the various measures to modernise farming because these changes have enabled targets set at higher organisational levels to be fulfilled. Meeting such targets is a major route towards promotion for officials across the country (O'Brien and Li 2017).

Because of its high-quality cherry production, the area of White Deer Plain around Duling and West Zhang slope villages is now known as Cherry Grove, attracting large numbers of tourists as well as bringing considerable economic income to the villagers (Song et al. 2017). The increased income is having tangible benefits, including improved housing conditions, higher standards of living, and improved quality of life for the villagers. The development of tourism has brought new investment to the villages to renovate roads, squares, parking lots, and other infrastructure. Farmers in the community have also empowered themselves due to the role of tourism, extending the rights of discourse, negotiation, and voting rights in village decision-making.

In terms of the notion of agricultural transition, there has been a wholesale move away from traditional grain production in the immediate hinterland of Xi'an. Despite the loss of areas of winter wheat production in Xi'an's hinterland, the city still derives much of its consumption of grain from domestic sources thanks to rising yields. Chinese grain consumption has tripled in the last 40 years, partly driven by more grains being fed to animals as meat consumption per capita rises, and so imports have also risen. Overall, between 2003 and 2017, China's food imports grew from just $14 billion to $104.6 billion, compared with $65 billion for food exports, as the country increasingly became part of the globalised food trade system (ChinaPower 2020). It is only when one drives further out across the Guanzhong Plain and reaches villages with poorer communications that there are more signs of village hollowing and the retention of grain production (Wang et al. 2019). However, even here there has been much recent government support to promote modernisation via grants for mechanisation and improved use of water as there can be problems of insufficient rainfall. In the immediate peri-urban fringe, horticulture has become dominant, partly reflecting individual entrepreneurship but supported by village committees and sponsored by municipal, district, and county administrations seeking to implement the strategic directions set out by central government. This has been a manifestation of a concerted focus on increased production, taken up enthusiastically by numerous individual farm households, but in most of the villages those households also derive significant income from tourist-related activities and from having some family members working off-farm. This gives rise to the economic multifunctionality evident across the Plain.

The link between farming and broader rural development is enshrined in the central government strategy to reduce rural poverty and raise rural incomes, which continue to lag behind urban levels. Support for agricultural modernisation has been strong, aimed at increasing farmers' incomes but also at meeting consumers' demands for higher-quality fresh produce. In White Deer Plain, modernisation extends beyond smallholder production not only into the spheres of marketing (e.g. e-commerce) and distribution (e.g. links to national supermarket and wholesaler chains) but also encouragement to develop larger production units via land leasing and entrepreneurial initiatives such as cooperative processing ventures and external capital for agribusiness development (e.g. wine production). Tourist development has seen government investment in improved rural infrastructure, large-scale attractions, and grants both to villages and individual households to improve visual appearance and facilities.

Much of the agricultural transition in White Deer Plain represents the intersection between government investment and the advance of market-driven capitalism. The former collectivised peasantry has transformed into commercially oriented small farmers with multiple income sources, some becoming members of the myriad of cooperatives that have developed or formed contracted linkages to agri-businesses (e.g. e-commerce, processing, and distribution firms). This differentiation of the peasantry has echoes of Kautsky's observations made over 100 years ago. The small farm sizes have limited the growth of household incomes so that most rural households remain on average substantially poorer than their urban counterparts, though income and enterprise diversification have helped to raise living standards. Concurrently, the wholesale switch to combined horticultural and tourism businesses has produced an economic multifunctionality that has been diffused across the Plain. These changes to the landscape are testament to an ongoing multifunctional agricultural transition.

References

Bao, J., K. Meng, and Q. Zhang. 2015. Rural urbanization led by tourism. *Geographical Research* 34 (8): 1422–1434 (in Chinese).

Bowler, I. R. 1986. Intensification, concentration and specialisation in agriculture: the case of the European Community. *Geography* 71: 14–24.

Chen, H., J. Marter-Kenyon, D. López-Carr, and X. Y. Liang. 2015. Land cover and landscape changes in Shaanxi Province during China's Grain for Green Program (2000–2010). *Environmental Monitoring and Assessment* 187 (10): 644.

Chen, J., J. Guan, J. B. Xu, and C. Clergeau. 2018. Constructing the green supply chain for rural tourism in China: perspective of front–backstage decoupling. *Sustainability* 10 (11): 4276.

China Labour Bulletin (CLB). 2020. Migrant workers and their children. *CLB*, May 11, 2020. https://clb.org.hk/content/migrant-workers-and-their-children

ChinaPower. 2020. *How is China feeding its population of 1.4 billion?* https://chinapower.csis.org/china-food-security/. Accessed August 25th, 2020.

Clegg, J. 2006. Rural cooperatives in China: policy and practice. *Journal of Small Business and Enterprise Development* 13 (2): 219–234.

De Clercq, M., A. Vats, and A. Biel. 2018. *Agriculture 4.0: The Future of Farming Technology*. World Government Summit, Oliver Wyman. www.worldgovernmentsummit.org

Fang, Y. G., and J. S. Liu. 2015. Diversified agriculture and rural development in China based on multifunction theory: beyond modernization paradigm. *Acta Geographica Sinica* 70 (2): 257–270.

Goodman, D., and M. Redclift. 1981. *From peasant to proletarian: capitalist development and agrarian transition*. Oxford: Basil Blackwell.

Griffiths, M. B., and J. Zeuthen. 2014. Bittersweet China: new discourses of hardship and social organisation. *Journal of Current Chinese Affairs* 43 (4): 143–174.

Gu, H., and C. Ryan. 2010. Hongcun, China—residents' perceptions of the impacts of tourism on a rural community: a mixed methods approach. *Journal of China Tourism Research* 6 (3): 216–243.

Huang, J., and S. Rozelle. 2018. China's 40 years of agricultural development and reform. In: *China's 40 years of reform and development*, edited by R. Garnaut, L. Song, and C. Fang, 487–506. Canberra: ANU Press.

Huang, Y. 2016. Understanding China's Belt & Road initiative: motivation, framework and assessment. *China Economic Review* 40: 314–321.

International Monetary Fund (IMF). 2020. World economic outlook database. https://www.imf.org/external/pubs/ft/weo/2020/01/weodata/

Ji, X, S. Rozell, J. Huang, L. Zhang, and T. Zhang. 2016. Are China's farms growing? *China & World Economy* 24 (1): 41–62.

Kautsky, K. 1988. *The agrarian question*. London: Zwan, two volumes.

Lenin, V. I. 1982. The differentiation of the peasantry. In: *Rural development: theories of peasant economy and agrarian change*, edited by J. C. Harriss, 130–138. London: Hutchinson.

Liu, Y. 2018. Introduction to land use and rural sustainability in China. *Land Use Policy* 74 (5): 1–4.

Liu, Y., Liu, Y., Chen, Y., and Long, H., 2010. The process and driving forces of rural hollowing in China under rapid urbanization. *Journal of Geographical Sciences* 20 (6): 876–888.

Lowder, S. K., J. Skoet, and T. Raney. 2016. The number, size, and distribution of farms, smallholder farms, and family farms worldwide. *World Development* 87: 16–29.

Lu, H, X. Hualin, H. Yafen, W. Zhilong, and Z. Xinmin. 2018. Assessing the impacts of land fragmentation and plot size on yields and costs: a translog production model and cost function approach. *Agricultural Systems* 161: 81–88.

Ministry of Agriculture and Rural Affairs of People's Republic of China (MARA). 2015. National Standard Release Conference for Beautiful Rural Construction (held in Beijing). http://www.moa.gov.cn/xw/zwdt/201505/t20150527_4620076.htm. Accessed May 25th, 2020.

Ministry of Culture and Tourism of the People's Republic of China (MCT). 2017. Notice of the National Tourism Administration on Printing and Distributing the Guidelines for the Creation of Global Tourism Demonstration Zones. http://zwgk.mct.gov.cn/auto255/201706/t20170612_832452.html?keywords. Accessed November 25th, 2019

Naughton, B., and K. S. Tsai (eds.). 2015. *State capitalism, institutional adaptation, and the Chinese miracle*. Cambridge: Cambridge University Press.

O'Brien, K. J., and L. Li. 2017. Selective policy implementation in rural China. In: *Critical readings on the Communist Party of China*, edited by K. E. Brodsgaard, 437–460. Leiden, Netherlands: Brill.

Park, C. H. 2014. Nongjiale tourism and contested space in rural China. *Modern China* 40 (5): 519–548.

Qian, Y, K. Song, T. Hu, and T. Ying. 2018. Environmental status of livestock and poultry sectors in China under current transformation stage. *Science of the Total Environment* 622: 702–709.

Robinson, G. M. 2004. *Geographies of agriculture: globalisation, restructuring and sustainability*. Harlow, UK: Pearson Education.

Robinson, G. M. 2018. Globalization of agriculture. *Annual Review of Resource Economics* 10: 133–160.

Robinson, G. M., and B. Song. 2019. Rural transformation: cherry growing on the Guanzhong Plain, China and the Adelaide Hills, South Australia. *Journal of Geographical Sciences* 29 (5): 675–701.

Schneider, M. 2017. Dragon head enterprises and the state of agribusiness in China. *Journal of Agrarian Change* 17 (1): 3–21.

Shepard, W. 2015. *Ghost cities of China: the story of cities without people in the world's most populated country*. London: Zed Books Ltd.

Song, B., G. M. Robinson, and Z. Zhou. 2017. Agricultural transformation and ecosystem services: a case study from Shanxi Province, China. *Habitat International* 69: 114–125.

Su, B. 2013. Developing rural tourism: the PAT program and 'Nong jia le' tourism in China. *International Journal of Tourism Research* 15 (6): 611–619.

Tan, M, G. M. Robinson, X. Li, and L. Xin. 2013. Spatial and temporal variability of farm size in China in context of rapid urbanization. *Chinese Geographical Science* 23 (5): 607–619.

The Pig Site. Market Report. 2020. The risk of African swine fever in the US and Australia. https://thepigsite.com/articles/market-report-the-risk-of-african-swine-fever-in-the-us-and-australia-june-2020

Wang, J, Y. Liu, and Y. Li. 2019. Ecological restoration under rural restructuring: a case study of Yan'an in China's loess plateau. *Land Use Policy* 87: 104087.

Wilson, G. A. 2007. *Multifunctional agriculture: a transition theory perspective*. Wallingford, UK and Cambridge, MA: CABI.

Wilson, G. A., 2001. From productivism to post-productivism… and back again? Exploring the (un) changed natural and mental landscapes of European agriculture. *Transactions of the Institute of British Geographers* 26 (1): 77–102.

Wilson, G. A., and R. J. F. Burton. 2015.Neo-productivist'agriculture: spatio-temporal versus structuralist perspectives. *Journal of Rural Studies* 38: 52–64.

Wu, X. 2014. The farmhouse joy (nongjiale) movement in China's ethnic minority villages. *The Asia Pacific Journal of Anthropology* 15 (2): 158–177.

Ye, J. 2018. Stayers in China's "hollowed-out" villages: a counter narrative on massive rural–urban migration. *Population, Space and Place* 24 (4): e2128.

Yu, J., and J. Wu. 2018. The sustainability of agricultural development in China: the agriculture–environment nexus. *Sustainability* 10 (6): 1776.

Zhang, J. 2019. Beyond the 'hidden agricultural revolution' and 'China's overseas land investment': main trends in China's agriculture and food sector. *Journal of Contemporary China* 28 (119): 746–762.

Zhao, X, H. Sun, B. Chen, X. Xia, and P. Li. 2019. China's rural human settlements: qualitative evaluation, quantitative analysis and policy implications. *Ecological Indicators* 105: 398–405.

Zhou, T, J. E. Vermaat, and X. Ke. 2019. Variability of agroecosystems and landscape service provision on the urban–rural fringe of Wuhan, Central China. *Urban Ecosystems* 22 (6): 1207–1214.

Zhou, X. (ed.). 2019. Across China: rural tourism draws villagers back to thrive. *Xinhuanet,* 8 (3). http://www.xinhuanet.com/english/2018-03/17/c_137046127.htm. Accessed March 8th, 2020.

Zinda, J. A. 2017. Tourism dynamos: selective commodification and developmental conservation in China's protected areas. *Geoforum* 78: 141–152.

5 A Checkered Pathway to Prosperity: The Institutional Challenges of Smallholder Tobacco Production in Zimbabwe

Tariro Kamuti

Introduction

This chapter is a description and critique of the inherent contradictions of smallholder tobacco production practices in Zimbabwe. Smallholder farmers are attracted to and engage in tobacco production as a livelihood strategy because it is lucrative. However, smallholder tobacco farming not only involves the clearing of forest land to make space to grow tobacco but also the cutting down of more trees elsewhere to provide fuelwood to cure the tobacco leaves. In the short-term, smallholder tobacco production earns good returns for the participating households but, in the long run, this practice is unsustainable because it leads to the gradual depletion and degradation of natural resources. These immediate gains constitute what I would refer to as a checkered pathway to prosperity, whereby various institutional processes which influence this increase in tobacco production are facilitating a rural transformation with negative aspects. These institutional processes and the subsequent rural changes caused by increased smallholder tobacco production exhibit inherent contradictions. Tobacco also referred to as the "golden leaf", has long been a significant foreign currency earner for the country; therefore, the government has supported this sector. The government has attributed current high tobacco production levels to the increased participation of smallholder farmers in the sector and has used this to hail the success of their land reform program and resultant black economic empowerment. However, government agencies responsible for conservation have been at loggerheads with smallholder tobacco farmers because their practices are leading to deforestation. Deforestation has devastating consequences which, together with climate change, have the potential to adversely affect poor, and hence vulnerable, people in the tobacco growing areas. Furthermore, tobacco has long received negative publicity because of its deleterious health effects, thus raising questions as to whether the crop should continue being produced and whether the government should continue to support this industry.

Tobacco production in Zimbabwe has increased sharply in the two decades since 2000, mainly from smallholder farmer output (Tobacco Industry Marketing Board 2014, 2015, 2018). These farmers are mainly located in rural areas under communal land tenure. They have been the beneficiaries of land redistribution since 1980, and especially of the fast-track land reform program since

DOI: 10.4324/9781003110095-6

the beginning of the new millennium. Communal land tenure in Zimbabwe involves common property rights whereby the land is collectively owned by villagers who have open access to natural resources like water, forest, pastures and land for settlement, arable agriculture and livestock production. The fast-track land reform program was an accelerated land redistribution initiative that started in the year 2000 and involved the forced takeover of previously white-owned land without compensation, subdividing the farms into smaller plots and redistributing the land to previously disadvantaged Zimbabweans from crowded areas which were historically designated as native reserves (Matondi 2012; Hanlon et al. 2013). The program is referred to as "fast-track" because it speeded up the process of land redistribution from the white minority, thus increasing the number of new beneficiaries who were resettled on that land. As a result of the fast-track land reform program, smallholder tobacco farmers numbers increased in the main tobacco-producing areas in the provinces of Manicaland, Mashonaland East, Mashonaland Central and Mashonaland West (Figure 5.1). Through the fast-track process, land tenure changed from the freehold title of the previously white-owned commercial farms to communal land tenure of the new resettlement areas. This transformation was greatest in the rural areas associated with

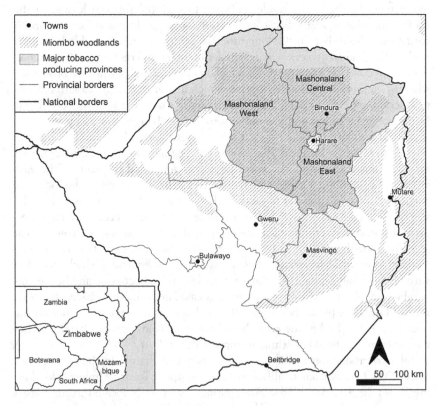

Figure 5.1 Miombo Woodlands and Tobacco Growing Areas.

Source: Author.

the increase in smallholder tobacco production resulting in the contradictions which are the focus of this chapter.

The increase in tobacco production is a result of Zimbabwe's agriculturally based economy and its predominantly rural population. The majority of Zimbabwe's population resides in rural areas and practices crop and livestock production for subsistence and the market. The recent increase in the number of smallholder farmers, some of whom grow tobacco as a cash crop, has enabled the mainstreaming of previously disadvantaged people into the wider economy. The tobacco sector was dominated by large-scale white commercial farmers before the year 2000. It has changed its character, and the majority of producers are now smallholder farmers (Manyanhaire and Kurangwa 2014, 1456; Munanga et al., 2014, 436). An anticipated result of this mainstreaming was an improvement in the living standards of rural households. At the same time, this shift has contributed to the transformation of the landscape. Tacit government support of the tobacco sector, coupled with a lack of capacity to enforce its key environmental protection policies, pose dilemmas for the governance and protection of the country's finite forest resources.

Smallholder tobacco production, as a result of its recent growth, is now a significant contributor to deforestation. This is particularly acute in the Miombo woodlands (Mangora 2012, 136; Njana et al., 2013, 128; Mandondo et al., 2014, 3). Miombo woodlands are natural forests that are composed of and mainly dominated by *Brachystegia* in combination with *Julbernardia* as well as *Isoberlinia* species (Sukume and Guveya 2003, 401–402). Miombo woodlands make up 41% of Zimbabwe's 39-million-hectare land area followed by 21% of the land area dominated by *Colophospermum mopane* (Chipika and Kowero 2000, 176; Sukume and Guveya 2003, 401; Mapedza 2007, 835) (Figure 5.1). The distribution of these forest resources mainly coincides with the major tobacco production areas in the three provinces of Mashonaland East, Mashonaland Central and Mashonaland West. Deforestation and the treatment of tobacco by using fuelwood derived from these tree species are further discussed below. This deforestation is in addition to that caused by the demand for fuelwood for domestic energy use by rural households that either do not have access to electricity (Whitlow 1980, 1; Sukume and Guveya 2003, 402) or who have electricity but face incessant outages. Key concerns which prompted scientific investigations of environmental degradation due to deforestation were raised both before the granting of Zimbabwean independence in 1980 and during the lead up to the fast-track land redistribution program in 2000 (Whitlow 1980, 1; Keeley and Scoones, 2000, 1). These concerns remain valid. The increase in the area under tobacco cultivation contributes to deforestation in three ways: by clearing vegetation to create space to grow the crop, by cutting down trees elsewhere to extract timber for the construction of tobacco barns and by further deforestation to obtain firewood which is used to cure the crop (Kagaruki 2010, 42; Mangora 2012, 135).

The governance of forest resources in the areas of tobacco production requires a degree of compromise if a viable livelihood strategy is to be achieved. The Zimbabwean government has supported the smallholder tobacco sector and has hailed its gains as a success of the fast-track land reform program which has empowered the many households who have been resettled on previously white-controlled land. The perceived economic benefits of tobacco farming in the fight against poverty in Africa have been used as an argument to stave off stringent control of the sector (Hu and Lee 2015, 46). The chapter next provides a theorization of the institutional processes that mediate the governance of forest resources by smallholder tobacco farmers and of how these result in the transformation of these rural areas. This is followed by a description of the methodological approach which relies on policy-related documents and tobacco production-related information available in the public domain. A snapshot of the smallholder tobacco sector in Zimbabwe follows, including a consideration of how the sector contributes to environmental degradation. A description of the institutional dynamics related to forest resources and tobacco production is followed by some concluding remarks on the checkered pathway to perceived success that is currently being followed by the smallholder tobacco producers.

A Theoretical Framework of Forest Institutions

Forests and the land which they occupy are resources that have the potential to drive rural development and thereby contribute to an improvement in peoples' livelihoods (Kowero 2003, 165; Njana et al. 2013, 124; Barrios et al. 2018, 1–2). Tobacco production can be seen as one way by which this natural resource can stimulate the economic growth of less developed countries (Anderson et al. 2006, 44). In the Zimbabwean case, communal land tenure coupled with the fast-track land redistribution program (Sukume and Guveya 2003, 402; Matondi 2012; Hanlon et al. 2013) have increased the levels of access to and stewardship over forest resources for tobacco production by low-income groups. The governance mechanisms mediating the access and use of those resources and the regulation of the tobacco sector, therefore, come into play. A governance perspective incorporates analysis of skewed power relations and resource ownership amongst the various actors in society as well as the differing perceptions of the nature of these resources (Olowu 2003, 503; Meadowcroft 2007, 306; Hirsch et al. 2010, 262). Governance is intricately linked to institutions (Crawford and Ostrom, 1995, 582) which are "systems of rules", whether formal or informal (Fleetwood 2008, 184). This chapter adopts Frances Cleaver's idea of institutional bricolage to explain the institutional processes behind the rural transformation currently being driven by the increase in smallholder tobacco production.

In adopting this concept, Cleaver (2012, 33) explains that "bricolage is a French word meaning to make creative and resourceful use of whatever materials are at hand, regardless of their original purpose". Bricolage is "'making do' with whatever is at hand" (Mair and Marti, 2009, 420). In this context, bricolage is understood as encompassing the adaptive processes through which people imbue the arrangements "of rules, traditions, norms and relationships" with "meaning and authority" (Cleaver, 2012, 34). In this process, old configurations are altered and new ones created, though these innovations must be related authoritatively to permissible ways of doing things if they are to be socially accepted. These modified arrangements refer to daily responses to varying situations. Cleaver asserts that "we are all bricoleurs" as the progress of our daily lives is made up of multitasking. However, institutional bricolage means more than just "making up and making do" (Cleaver, 2012, 34). Processes of bricolage will not simply work; rather institutions need to "be legitimised and imbued with authority" for these processes to be adopted "and to endure over time and space" (Cleaver, 2012, 34). Cleaver points out "that the concept of institutional bricolage offers a way of analysing and understanding just how institutions are socially formed and practised" in these instances (Cleaver, 2012, 35). Applying the concept of institutional bricolage at the level of government, the various departments and stakeholders possessing oversight of smallholder tobacco farming constitute the bricoleurs (that is the stakeholders, players or actors).

Institutions play a critical role in, for example, overseeing changes in adaptation strategies in the face of climate change to enhance sustainability (Berman et al. 2012, 93). Concerning smallholder tobacco production in Zimbabwe, these institutions promulgate policies, laws and regulations and create the entities that are put in place to formulate, modify and implement these policies, laws and regulations. Such entities are numerous but the Ministry of Environment, Water and Climate, Ministry of Lands, Water and Agriculture, the Forestry Commission, the Environmental Management Agency (EMA), the Tobacco Industry and Marketing Board (TIMB), District Rural Councils (representing local government), farmer associations, tobacco companies, non-governmental organizations and smallholder farming communities, down to their household and individual levels, are among the most important. The government of Zimbabwe, through its various ministries and agencies, is ultimately responsible for providing and enforcing environmental and tobacco-related policies, laws and regulations from the national to the local scale. Tobacco companies buy tobacco from the farmers for processing into various products for local consumption and export. Non-governmental organizations deal with various issues connected to the tobacco industry such as environmental protection, the welfare of workers and the anti-tobacco lobby among others. Some agricultural policies directly link agricultural production to deforestation in the communal areas of Zimbabwe, while others deal with population, livestock production, poverty and income (Chipika and Kowero 2000, 178).

To date, there has been a dearth of studies on the direct impacts of tobacco production on the livelihoods of the low-income groups who comprise the majority of tobacco producers in Zimbabwe. It is within this context that this chapter seeks to explore the dynamics around the challenges posed by the tobacco sector in terms of the major governance issues, especially given the controversies surrounding the health issues associated with the crop.

The concept of institutional bricolage posits that the institutions dealing with natural resources management are not necessarily well designed for that purpose as they are often vague, they have multiple functions, they are ever-changing and are less susceptible to intentional crafting (Cleaver and Franks 2005, 17). Drawing on data from key statutory institutions, this study argues that, while there have been improvements in household income and a subsequent rise in living standards from tobacco sales (Masvongo et al. 2013, 300), the sector is straining forest resources to an extent that requires a radical shift in institutional arrangements. Several measures have been put in place to stem the deforestation associated with small-scale tobacco production though, to date, these have brought mixed results. Implementation of these measures by the various stakeholders in the sector has not been uniform. Policymaking is complex and not straightforward (Keeley and Scoones 2000, 3); hence, the institutions responsible for natural resource management are not necessarily path-dependent in accordance with design principles, but rather they have undergone a long process of institutional bricolage and development (Kowero et al. 2003, 166; Cleaver 2012). For example, it is argued here that no one intentionally attempts to destroy forests, but this is an unavoidable result of the efforts of smallholder farmers to eke out a living (Mangora 2012, 138). One is then drawn to ask if this is the case for the smallholder tobacco production sector in Zimbabwe.

Methodological Approach

This chapter is based on policy-related matters. Hence, it refers to policy materials, such as the Constitution of Zimbabwe Amendment (No. 20) Act of 2013 and a string of legislation devised by the various government agencies (Government of Zimbabwe, 2013). The Environmental Management Act (Chapter 20:27) deals with environmental protection in general. Specific legislation dealing with forest resources includes the Forest Act [Chapter 19:05], the Communal Land Forest Produce Act [Chapter 19:04] and the statutory instrument: Forest (Control of Firewood, Timber and Forest Produce) Regulations of 2012. This study also refers to secondary data, such as descriptive statistics (for example, from the Tobacco Industry Information and Marketing Board) on the smallholder tobacco sector in Zimbabwe, together with material on the production, marketing, consumption and effects of tobacco. These sources are utilized to unravel the interactions of the key players in the smallholder tobacco sector and the formal and informal regulations applying to them to understand their contributions to rural transformation in this region of Zimbabwe. These enduring interactions determine both the condition of those forest resources that are impacted by tobacco production and the livelihoods and living standards of the smallholder tobacco farmers and their families. The key players are the government agencies responsible for the regulation of natural resources and tobacco production, the tobacco companies that purchase the tobacco, the tobacco farmers who produce the crop and the farmers' organizations that represent the interests of the smallholders.

The chapter draws on discourse analysis to assist in connecting the main ideas of the theoretical framework of institutional bricolage to the issues arising from the growth of the smallholder tobacco sector in Zimbabwe. This is based on considerations of how participant households in the rural areas structure and process their agricultural activities as a product of their perceptions and evaluations of their livelihood challenges which, in turn, are formed in the contexts of their assessment of their political, social, cultural and economic dispositions and milieux. This notion fuses the "discourses of rurality [which are] produced and reproduced by ordinary people in their everyday lives" with their quest for better livelihoods (Woods 2005, 13). Each of the actors (that is, the "bricoleurs", as explained in the theoretical framework) in the smallholder tobacco production arena gives meaning to the world as they see it, for example, in terms of how they perceive shifting market forces and the regulation of the sector since these directly affect their interests and subsequent operations. These social constructions of rurality by the smallholder tobacco farmers help to explain the nature of their participation in the transformations witnessed in these rural areas.

Smallholder Tobacco Production in Zimbabwe

Overall, there has been an increase in the number of smallholder flue-cured tobacco farmers in Zimbabwe from 8,537 growers in 2000 (Tobacco Industry Marketing Board 2014) to 140,895 in 2018 (Tobacco Industry Marketing Board 2018, 26). The fast-track land reform program has benefited a sizeable proportion of small-scale commercial farmers who were mainly classified as A1 and A2 under the models of land redistribution. The A1 farmers already had small landholdings, while the A2 farmers were given large pieces of what was formerly white settler farmland upon which to practice commercial agriculture (Mkodzongi and Lawrence 2019, 1). The communal area farmers occupy marginal land in what was previously termed reserves, namely those areas to which African people were restricted during the colonial period. So, the A1 and A2 farmers are mainly people who moved from the communal areas as part of efforts to decongest areas in which the land was already less productive. However, further deforestation, driven by the upsurge in tobacco production, is worsening the degradation of landscapes in the communal areas. The high numbers of registered growers in the 2017/18 growing season (Table 5.1) as compared to previous seasons is instructive in this respect. The three northern provinces of Mashonaland East, Mashonaland Central and Mashonaland West provide the bulk of tobacco produced in the country (Figure 5.1). In Zimbabwe, as in Malawi, this increase in tobacco production has been brought about by an increase in the area under tobacco cultivation rather than by improvements in productivity (Mandondo et al., 2014, 1; Tobacco Industry Marketing Board 2014, 31).

Flue-cured tobacco sales take place through auction sales and contract sales. Out of the 87,166 growers who sold their produce in 2014, about 38,023 of them marketed their produce at the auction floors while 49,143 growers were under

Table 5.1 Registered Tobacco Growers According to Sector

Sector	2012/13	2013/14	2014/15	2016/17	2017/18
A1	38,020	37,805	35,532	38,103	52,918
A2	8,218	11,720	9,021	7,658	9,190
Communal area	37,494	48,292	45,617	46,621	70,551
Small-scale commercial	8,546	8,639	7,465	6,545	8,236
Total	**91,278**	**106,456**	**97,635**	**98,927**	**140,895**

Source: TIMB 2014, 2015, 2018.

contract (Tobacco Industry Marketing Board 2014, 9). Contract growers are assured of a fixed selling price for their produce when they start to plant the crop, while those growers who trade on the auction market experience varying prices in any particular season. This brings stability to the incomes of contract growers as compared to those who trade their produce on the auction market. For example, auction sales for the year 2014 were 50.7 million kilograms at an average seasonal auction price of US$2.69/kilogram, while contract sales were 165.5 million kilograms at a seasonal average contract price of US$3.32/kilograms (Tobacco Industry Marketing Board 2014, 24, 25). For the year 2015, auction sales comprised 46.7 million kilograms at a seasonal average auction price of US$2.43/kilogram compared to 152.3 million kilograms of contract sales at a seasonal average contract price of US$3.11/kilogram (TIMB, 2015, 5). The Tobacco Industry Marketing Board (2014, 10, 12) noted a decline in auction sales and attributed this to firming contract sales prices that went as high as US$6.20/kilogram. In the Mazowe District (Mashonaland Central Province), contract farmers produced more tobacco than non-contract farmers due to "access to inputs, extension services and finance" though there was no significant variation in the prices that each class of farmers obtained from the market (Moyo 2014, iv). Such an increase in tobacco production through contract arrangements has the potential to improve the livelihoods of the farmers through a boost in productivity due to increased efficiency (Moyo 2014, iv). Major variations in tobacco production are a characteristic of the sector as illustrated by the fluctuations that have occurred between 2000 and 2018 (Table 5.2). Tobacco as a cash crop is

Table 5.2 Flue-Cured Tobacco Production 2000–2018

Year	Growers	Area (ha)	Mass sold (kg)	Average price (US$/kg)	Average yield (kg/ha)	Gross value (US$)
2018	140,895	133,000	252,603,251	2.92	1,899	737,603,251.00
2017	98,927	110,816	188,920,313	2.96	1,705	559,077,353.00
2016	81,801	102,537	202,275,688	2.95	1,972	595,927,523.00
2015	97,616	104,662	198,954,849	2.95	1,900	586,544,231.00
2014	106,372	102,537	216,196,683	3.17	2,108	685,244,013.00
2013	78,756	88,627	166,572,097	3.67	1,879	612,135,672.00
2012	60,047	76,359	144,565,253	3.65	1,893	527,805,943.00
2011	56,656	78,415	132,431,905	2.73	1,689	361,448,679.00
2010	51,685	67,054	123,503,681	2.88	1,842	355,572,326.00
2009	29,018	62,737	58,570,652	2.98	934	174,457,761.00
2008	35,094	61,622	48,775,178	3.21	792	156,663,816.00
2007	26,412	54,551	73,039,015	2.32	1,339	169,159,675.00
2006	20,565	58,808	55,466,689	2.00	943	–
2005	31,761	57,511	73,376,990	1.61	1,300	–
2004	21,882	44,025	68,901,129	2.00	1,565	–
2003	20,513	49,571	81,806,414	2.25	1,673	–
2002	14,353	74,295	165,835,001	2.27	2,213	–
2001	7,937	76,017	202,535,209	1.75	2,664	–
2000	8,537	84,857	236,946,295	1.69	2,792	–

Source: TIMB 2014, 2015, 2018.

dependent on global processes and the changes in the pricing of the product on the international market have a bearing on tobacco earnings at a local level and ultimately on household incomes. Government regulations on earnings from tobacco exports also have a role to play. So, the livelihood of a village tobacco farmer is influenced by the shifts in global markets and this, in turn, affects the extent to which local farmers engage in tobacco production in each season.

Deforestation and Environmental Concerns due to Tobacco Production

The tree species that constitute the Miombo woodlands are in high demand as fuelwood for curing tobacco (Mangora 2012, 135) due to their prevalence across the country (as shown in Figure 5.1) and also their high energy output (Chipika and Kowero 2000, 176; Manyanhaire and Kurangwa 2014, 1463). Information on the effects of tobacco production on forest resources is available from an attempt by Manyanhaire and Kurangwa (2014) to estimate the quantity of firewood needed to cure tobacco in Zimbabwe. They noted an upsurge in the use of woodland resources that were directly related to the increase in tobacco production across the country. Their conclusion was based on calculations of wood consumption using a "woodland clearance factor of 0.6 hectares per one hectare of cured tobacco and wood consumption factor of 14 kilograms per one kilogram of cured tobacco" (Manyanhaire and Kurangwa 2014, 1459). An estimated 300,000 hectares of indigenous woodlands were cleared by smallholder farmers per year during the period 2010 to 2012 (Musoni et al. 2013, 609). This increase in wood consumption for tobacco curing was occurring despite the existence of tobacco industry support mechanisms for the provision of coal as an alternative source of energy for the same purpose (Manyanhaire and Kurangwa 2014, 1456). The conventional wood-fueled tobacco barn (Figure 5.2) used by the majority of smallholder farmers (Musoni et al., 2013, 609) is also considered to be less efficient than other curing methods (Manyanhaire and Kurangwa 2014, 1456; Munanga et al. 2014, 436). There is, therefore, a need to adopt new initiatives such as the Rocket barn which consumes 47%–50% of fuelwood required by a conventional barn to cure the same quantity of tobacco (Munanga et al. 2014, 436). A new design of barn which is 54.7% and 74.2% more effective than the Rocket barn and the conventional barn, respectively, is recommended by Musoni et al. (2013, 609).

Globally, there is a skewed distribution of tobacco income due to capital flows toward the developed countries where multinational tobacco companies such as British American Tobacco and Philip Morris (Riquinho and Hennington 2012, 1588) are headquartered. British American Tobacco and Philip Morris have been criticized for using superficial public relations exercises, such as allegedly "green supply chains" that promote discourses of tobacco production as "environmentally and socially friendly", rather than taking genuine initiatives to stem social and environmental abuses, such as child labor and deforestation, in the developing world (Otañez and Glantz 2011, 403). Smallholder tobacco agriculture in Zimbabwe relies on family labor, and the use of child labor is rife. Power relations

Figure 5.2 A Conventional Tobacco Barn and Logs of Firewood from Indigenous Trees.
Source: Author.

are skewed in favor of a few influential individuals and corporate entities and against local communities who bear the brunt of the burden and costs of producing tobacco (Loker 2005, 311; Ribot et al. 2010, 39; Hu and Lee 2015, 43). In addition to exploitative labor practices against women and children, smallholder tobacco production exposes those involved to health hazards related to contact with pesticides that damage the environment, and to the inhalation of carbon monoxide and other poisonous substances in the curing of tobacco (Loker 2005, 321; Kagaruki 2010, 41; Otañez and Glantz 2011, 403). These situations, coupled with the increased consumption of tobacco in the developing world, have negative implications on the quality of life in tobacco-growing communities. For instance, the study of the rise and fall of flue-cured tobacco in the Copán Valley of Honduras highlights "the importance of paying close attention to the interactions among local agroecology, social conditions, the natural environment, and the impulses emanating from the broader political economy and how these resonate at the local level" (Loker 2005, 324). Any attempt to integrate environmental targets, livelihood needs and democratic goals are challenging but, in this context, local people, by shouldering the burden of nurturing forest resources, would deny themselves the opportunity to earn a livelihood. Conversely, most positive spin-offs from smallholder tobacco production accrue to international elites (Ribot et al. 2010, 40).

The top producers of tobacco in the world are China, Brazil, India, the USA and Malawi but, in terms of exports, Brazil, Zimbabwe and Malawi, respectively, dominate world trade (Riquinho and Hennington 2012, 1591). Tobacco constitutes about 50% of Zimbabwe's agricultural production, 35% of its exports and around 10% of its Gross Domestic Product (Musoni et al. 2013, 609). A study of the financial sustainability of small-scale tobacco output and its effects on forest resources in Hurungwe District (Mashonaland West Province), found that farmers benefit significantly from the proceeds of tobacco sales when the expenses for firewood are not factored in (Chivuraise 2011, iv). There was therefore a suggestion "to put in place policies that govern natural forest depletion such as gum plantations, subsidizing [the] price of coal and introduc[ing] fees, penalties or taxes [for] the offenders" (Chivuraise 2011, iv). In Malawi, agroforestry offered a viable opportunity for carbon sequestration in the drive toward reducing emissions from deforestation and forest degradation (Thangata and Hildebrand 2012, 172). In my view, the adoption of such strategies would help to reduce the carbon footprint of tobacco production, thereby facilitating the sustainable utilization of forest resources and enhancing the livelihoods of smallholder farmers.

By increasing their output, smallholder tobacco farmers have capitalized on opportunities offered to them through their own networks and through cooperation with the Ministry of Environment, Water and Climate, the Ministry of Lands, Water and Agriculture, the Forestry Commission, the EMA, the TIMB, and with District Rural Councils (representing local government), essentially therefore with like-minded people in the sector. Thus, institutional bricolage happens in a broader context than that of the discernible configurations of the official resource management, i.e. the statutory bodies. Since 2000, Zimbabwe has experienced extreme political instability and an economic crisis that has minimized the role of the state in service provision though it has maintained its grip on authority (Cleaver 2012). This desperate situation in Zimbabwe nevertheless offered an opportunity for the smallholder tobacco sector to expand and to act as a coping mechanism for rural households faced with few if any other viable livelihood alternatives. The opportunities offered by the cultivation of tobacco attracted the less disadvantaged sections of Zimbabwean society since they ignored, at least in the short term, the environmental and health costs of the crop. Hence, the results of the growth of this industry have been mixed for the participating households. The results of this expansion for the participating institutions and for the natural environment have also varied as the patchwork involving the old and the new institutions has struggled to adapt to this situation of change and crisis (Cleaver 2012).

Institutional Dynamics Related to Forest Resources and Tobacco Production

Section 73 of the Constitution of Zimbabwe refers to the environmental rights of citizens and the role of the State in that regard (Government of Zimbabwe

2013, 37). Notably, in Section 73(1)(b), the Constitution states that it is the right of every person

> to have the environment protected for the benefit of present and future generations, through reasonable legislative and other measures that ... (ii) promote conservation; and (iii) secure ecologically sustainable development and use of natural resources while promoting economic and social development.
>
> (Government of Zimbabwe 2013, 37)

Section 73(2) goes on to state that "the state must take reasonable legislative and other measures, within the limits of the resources available to it, to achieve the progressive realization of the rights set out in this section" (Government of Zimbabwe 2013, 37). Broadly Section 73 of the Constitution, as the supreme law of Zimbabwe, sets the overarching framework which defines both the institutions of governance and the "rules, norms, and shared strategies" that "are constituted and reconstituted by human interaction in frequently occurring or repetitive situations" (Crawford and Ostrom 1995, 582).

These guidelines from the Constitution direct the Ministry of Environment, Water and Climate, which oversees state instrumentalities such as the Environmental Management Agency and the Forestry Commission. For example, the Forestry Commission states that it "contributes to national socio-economic development through regulation and capacity enhancement in the utilization and management of forest resources"[1] by applying the Forest Act (Chapter 19:05 of 1999) and the Communal Lands Forest Produce Act (Chapter 20 of 1987). In line with changing situations, as noted above in the arguments for institutional bricolage, the Forest Act (Chapter 19:05 of 1999) has undergone many amendments since its enactment on the 9th of December 1949 (Government of Zimbabwe 1999). Mapedza (2007, 851) points out that some of the changes to these forest regulations fundamentally limited indigenous peoples' access to and control of forest resources (see also Mataya et al. 2003, 110). The most recent amendments to this Act were made in 1999, just before the commencement of the much-vaunted fast track land reform program in 2000, and the provisions of the Act as it stands may well be inadequate to deal with the increased rate of deforestation that has accompanied the fast-track land reform program and the subsequent increase in the production of tobacco.

However, the Communal Lands Forest Produce Act (Chapter 20 of 1987) was specifically intended to address issues related to forest resources in communal areas where tobacco production has drastically increased since 2000. Furthermore, the government has created a new statutory body, the Environmental Management Agency, through the Environmental Management Act (Chapter 20:27) of 2002.[2] The Environmental Management Act (Chapter 20:27) makes the Environmental Management Agency's mandate broader than those of other statutory bodies dealing with the environment by stating, in Section 3(2), that "if any other law is in conflict or inconsistent with this Act, this Act shall prevail" (Government of Zimbabwe 2002, 8). The Environmental Management Agency was established because of the need to address all the environmental challenges (including deforestation due to tobacco production) facing the country. It was therefore given a broad mandate. However, if one looks closely at the role of the Environmental Management Agency concerning forest resources, there is an overlap between its responsibilities and those of the Forestry Commission. Both agencies have the legal authority to reduce deforestation. However, due to a shortage of resources resulting from the economic crisis, neither statutory body has been able to fulfil its role.

The tacit encouragement of tobacco farming by the government through the Ministry of Lands, Water and Agriculture as a means of economic emancipation clashes with the ecological mandate of the Ministry of Environment, Water and Climate. This presents a serious challenge to the government's overall mission to "secure ecologically sustainable development and use of natural resources while promoting economic and social development" (Government of Zimbabwe 2013, 37). Thus, smallholder farmers will continue to grow tobacco as a livelihood strategy as long as they can make a profit, despite the looming and increasingly apparent negative environmental consequences of this trend. Thus, while in some societies, rural transformations have involved moves toward the adoption of post-productive strategies, in Zimbabwe the entrenchment of primary extractive activities through smallholder tobacco production relies heavily on the depletion of its forest resources.

Institutions are critical in shaping the use and governance of environmental resources, especially at the local level, and this adds a further level of complexity under multilevel governance systems (Slavíková et al. 2010, 1369). In Zimbabwe, negotiations over access to forest resources have been largely left to communities at the local level (Keeley and Scoones 2000, 33). However, these local institutions are weak, government capacity is stretched, and the overlapping and sometimes conflicting roles of its agencies in the monitoring and control of access and utilization of forest resources further inhibit their effectiveness (Mataya et al. 2003, 97). Hence, massive deforestation has proceeded at an alarmingly high rate in a way that threatens both the long-term viability of the tobacco sector and the sustainability of the natural forest resources. The fast-track land reform program was a culmination of political manoeuvring by the ruling party which has had numerous consequences, one of which was the distribution of previously large-scale commercially owned land to members of low-income groups. This shifted the market economics of private property rights resulting in the devastating macro-economic instability (hyper-inflation) which caused the government to abandon the Zimbabwean dollar as the official currency.

Nevertheless, the fast-track land reform program also signalled a break in the "continuity between the colonial and immediate post-colonial periods in terms of the way policy has operated" (Keeley and Scoones 2000, 22), namely the restitution of land that was confiscated from the indigenous population (Mapedza 2007, 851). Forest policies have, for a long period, regarded indigenous peoples' agricultural practices as being detrimental to the environment (Keeley and Scoones 2000, 5; Kowero et al. 2003, 169; Kwashirai 2007; Mapedza 2007, 842). Hence, the ruling party's policy of decolonizing the economy as promised in the 2000 parliamentary and 2002 presidential elections, bolstered and legitimized its radical stance of confiscating land from the white commercial farmers thereby empowering the majority African population as the rightful owners of the country's natural resources. Certainly, not all sections of society agreed with the government's stance of steering the Zimbabwean economy through command tactics. Indeed, Slavíková et al. (2010, 1370) argued that the core focus of the state should be on establishing and safeguarding robust institutions for sustainable governance.

Policy scholars and policymakers should not assume that all parts of the world follow linear and predictable social, economic and ecological trajectories (Keeley and Scoones 2000, 3; Duit and Galaz 2008, 311). In Zimbabwe's case, its economic policies have permeated other sectors, particularly agriculture and the environment, with unpredicted and negative results concerning forest resources. Policies that seem good in theory frequently prove difficult to implement. Hence, there is an ongoing need to restructure or replace institutions after carefully monitoring specific outcomes and paying attention to the ecological, economic and social dimensions of the issues at stake, ideally with the use of transparent social dialogue (Slavíková et al. 2010, 1370). The changes in the institutional arrangements relating to smallholder tobacco production in Zimbabwe can be explained through the lens of institutional bricolage. However, this does not alter the fact that smallholder tobacco farmers are satisfying their immediate livelihood needs in a manner that is unsustainable.

Conclusion

Smallholder tobacco farmers in Zimbabwe have found an immediate way to extricate themselves from poverty. However, with forest resources dwindling at their current high rate, tobacco production will become more costly in future, thus affecting its profitability and, ultimately, its sustainability (Masvongo et al. 2013, 300). If this outcome is to be avoided, it will be necessary to restructure the existing environmental management institutions or to develop new ones (Slavíková et al. 2010, 1369). It is the nature of institutions that they need to evolve continuously to keep pace with new developments such as, in this case, the increase in smallholder tobacco production. Ideally, the overlaps in the remits of the institutions governing land, forest and other resources should be reduced while the capacity of each institution to fulfil its core functions should be strengthened.

Whitlow's (1980, v) call for the implementation of strategies that reduce deforestation to be integrated within rural development initiatives is still relevant today. Different forms of partnerships in the management of common property resources (Nemarundwe 2001, 1) need to be fostered to increase the level of capital, human and financial resources that can be possibly directed toward protecting and conserving indigenous forest resources. All these measures are in line with the long-drawn-out processes of the development of appropriate institutional mechanisms that mediate access to and utilization of indigenous forest resources for various purposes, including the curing of tobacco. These measures should also be adopted in conjunction with the ambit of the state's role as pre-scribed in Section 73(2) of the Zimbabwean Constitution which provides for reasonable legislative and relevant measures within the limits of the resources at its disposal. Otherwise, as long as rural households continue to benefit from producing tobacco through exploiting indigenous forest resources without any drastic consequences, either through immediate negative environmental reper-cussions or through self-initiated sanctions and external pressure (the checkered path), the depletion of these resources will become severe.

Notes

1 See "Welcome to The Forestry Commission" Available at: http://www.forestry.co.zw/ Accessed: 29/09/2016.
2 See "Vision, Mission Statement & Core Values" Available at: http://www.ema.co.zw/ index.php/2014-06-12-03-49-33/mission-vision-values.html Accessed: 29/09/2016.

References

Anderson, J., C. Benjamin, B. Campbell, and D. Tiveau. 2006. Forests, Poverty and Equity in Africa: New Perspectives on Policy and Practice. *International Forestry Review* 8 (1): 44–53.

Barrios, E., V. Valencia, M. Jonsson, A. Brauman, K. Hairiah, P. E. Mortimer, and S. Okubo. 2018. Contribution of Trees to the Conservation of Biodiversity and Ecosystem Services in Agricultural Landscapes. *International Journal of Biodiversity Science, Ecosystem Services & Management* 14 (1): 1–16.

Berman, R., C. Quinn, and J. Paavola. 2012. The Role of Institutions in the Transformation of Coping Capacity to Sustainable Adaptive Capacity. *Environmental Development* 2: 86–100.

Chipika J. T., and G. Kowero. 2000. Deforestation of Woodlands in Communal Areas of Zimbabwe: Is It Due to Agricultural Policies? *Agriculture, Ecosystems and Environment* 79: 175–185.

Chivuraise, C. 2011. The Economics of Smallholder Tobacco Production and Implications of Tobacco Growing on Deforestation in Hurungwe District of Zimbabwe. MSc disser-tation, University of Zimbabwe.

Cleaver, F. 2012. *Development through Bricolage: Rethinking Institutions for Natural Resources Management.* London: Routledge.

Cleaver, F. and T. Franks. 2005. How Institutions Elude Design: River Basin Management and Sustainable Livelihoods. Bradford Centre for International Development (BCID) Research Paper No. 12. Bradford: University of Bradford.

Crawford, S. E. S., and E. Ostrom. 1995. A Grammar of Institutions. *American Political Science Review* 89 (3): 582–600.

Duit, A., and V. Galaz. 2008. Governance and Complexity—Emerging Issues for Governance Theory. *Governance: An International Journal of Policy, Administration, and Institutions* 21 (3): 311–335.

Fleetwood, S. 2008. Structure, Institution, Agency, Habit, and Reflexive Deliberation. *Journal of Institutional Economics* 4 (2): 183–203.

Government of Zimbabwe. 1999. *Forest Act (Chapter 19:05 of 1999).* Harare: Government Printer.

Government of Zimbabwe. 2002. *Environmental Management Act (Chapter 20:27).* Harare: Government Printer.

Government of Zimbabwe. 2013. *Constitution of Zimbabwe Amendment (No. 20) Act, 2013.* Harare: Government Printer.

Hanlon, J., J. Manjengwa, and T. Smart. 2013. *Zimbabwe Takes Back Its Land.* Johannesburg: Jacana Media.

Hirsch, P. D., W. M. Adams, J. P. Brosius, A. Zia, N. Bariola, and J. L. Dammert. 2010. Acknowledging Conservation Trade-Offs and Embracing Complexity. *Conservation Biology* 25 (2): 259–264.

Hu, T.-W., and A. H. Lee. 2015. Commentary: Tobacco Control and Tobacco Farming in African Countries. *Journal of Public Health Policy* 36: 41–51.

Kagaruki, L. K. 2010. Community-Based Advocacy Opportunities for Tobacco Control: Experience from Tanzania. *Global Health Promotion* 17 (2): 41–44.

Keeley, J., and I. Scoones. 2000. Environmental policymaking in Zimbabwe: Discourses, Science and Politics. Institute of Development Studies Working Paper 116. Brighton: University of Sussex.

Kowero, G. 2003. The Challenge to Natural Forest Management in Sub-Saharan Africa Rural Development: Experiences from the Miombo Woodlands of Southern Africa. In *Policies and Governance Structures in Woodlands of Southern Africa*, ed. G. Kowero, B. M. Campbell, and U. R. Sumaila, 1–8. Jakarta: Center for International Forestry Research.

Kowero, G., A. R. S. Kaoneka, I. Nhantumbo, P. Gondo, and C. B. L. Jumbe. 2003. Forest Policies in Malawi, Mozambique, Tanzania and Zimbabwe. In *Policies and Governance Structures in Woodlands of Southern Africa*, ed. G. Kowero, B. M. Campbell, and U. R. Sumaila, 165–186. Jakarta: Center for International Forestry Research.

Kwashirai, V. C. 2007. Indigenous Management of Teak Woodland in Zimbabwe, 1850–1900. *Journal of Historical Geography*, 33 (4): 816–832.

Loker, W. M. 2005. The Rise and Fall of Flue-Cured Tobacco in the Copán Valley and Its Environmental and Social Consequences. *Human Ecology* 33: 299–327.

Mair, J., and I. Marti. 2009. Entrepreneurship in and Around Institutional Voids: A Case Study from Bangladesh. *Journal of Business Venturing* 24: 419–435.

Mandondo, A., L. German, H. Utila, and U. M. Nthenda. 2014. Assessing Societal Benefits and Trade-Offs of Tobacco in the Miombo Woodlands of Malawi. *Human Ecology* 42: 1–19.

Mangora, M. M. 2012. Shifting Cultivation, Wood Use and Deforestation Attributes of Tobacco Farming in Urambo District, Tanzania. *Current Research Journal of Social Sciences* 4 (2): 135–140.

Manyanhaire, I. O., and W. Kurangwa. 2014. Estimation of the Impact of Tobacco Curing on Wood Resources in Zimbabwe. *International Journal of Development and Sustainability* 3 (7): 1455–1467.

Mapedza, E. 2007. Forestry Policy in Colonial and Postcolonial Zimbabwe: Continuity and Change. *Journal of Historical Geography*, 33 (4): 833–851.

Masvongo, J., J. Mutambara, and A. Zvinavashe. 2013. Viability of Tobacco Production under Smallholder Farming Sector in Mount Darwin District, Zimbabwe. *Journal of Development and Agricultural Economics* 5 (8): 295–301.

Mataya, C., P. Gondo, and G. Kowero. 2003. Evolution of Land Policies and Legislation in Malawi and Zimbabwe: Implications for Forestry Development. In *Policies and Governance Structures in Woodlands of Southern Africa*, ed. G. Kowero, B. M. Campbell, and U. R. Sumaila, 92–112. Jakarta: Center for International Forestry Research.

Matondi, P. B. 2012. *Zimbabwe's Fast Land Reform.* London: Zed Books.

Meadowcroft, J. 2007. Who is in Charge Here? Governance for Sustainable Development in a Complex World. *Journal of Environmental Policy and Planning* 9 (3-4): 299–314.

Moyo, M. 2014. Effectiveness of a Contract Farming Arrangement: A Case Study of Tobacco Farmers in Mazowe District in Zimbabwe. MPhil dissertation, University of Stellenbosch.

Mkodzongi, G., and P. Lawrence. 2019. The Fast-Track Land Reform and Agrarian Change in Zimbabwe. *Review of African Political Economy* 46 (159): 1–13.

Munanga, W., C. Kufazvinei, F. Mugabe, and E. Svotwa. 2014. Evaluation of the Curing Efficiency of the Rocket Barn in Zimbabwe. *International Journal of Agriculture Innovations and Research* 3 (2): 436–441.

Musoni, S., R. Nazare, E. Manzungu, and B. Chekenya. 2013. Redesign of Commonly Used Tobacco Curing Barns in Zimbabwe for Increased Energy Efficiency. *International Journal of Engineering Science and Technology* 5 (3): 609–617.

Nemarundwe, N. 2001. Institutional Collaboration and Shared Learning for Forest Management in Chivi District, Zimbabwe. IES Working Paper 16, Institute of Environmental Studies, University of Zimbabwe.

Njana, M. A., G. C. Kajembe, and R. E. Malimbwi. 2013. Are Miombo Woodlands Vital to Livelihoods of Rural Households? Evidence from Urumwa and Surrounding Communities, Tabora, Tanzania. *Forests, Trees and Livelihoods* 22 (2): 124–140.

Olowu, D. 2003. Challenge of Multi-Level Governance in Developing Countries and Possible GIS Applications. *Habitat International* 27: 501–522.

Otañez, M., and S. A. Glantz. 2011. Social Responsibility in Tobacco Production? Tobacco Companies' Use of Green Supply Chains to Obscure the Real Costs of Tobacco Farming. *Tobacco Control* 20 (6): 403–411.

Ribot, J. C., J. F. Lund, and T. Treue. 2010. Democratic Decentralization in Sub-Saharan Africa: Its Contribution to Forest Management, Livelihoods, and Enfranchisement. *Environmental Conservation*, 37: 35–44.

Riquinho, D. L., and E. A. Hennington. 2012. Health Environment and Working Conditions in Tobacco Cultivation. *Ciênc and Saúde Coletiva* 17 (6): 1587–1600.

Slavíková, L., T. Kluvánková-Oravská, and J. Jílková. 2010. Bridging Theories on Environmental Governance Insights from Free-Market Approaches and Institutional Ecological Economics Perspectives. *Ecological Economics* 69: 1368–1372.

Sukume, C., and E. Guveya. 2003. A System Dynamics Model for Management of Miombo Woodlands in Two Communal Areas of Zimbabwe. In *Policies and Governance Structures in Woodlands of Southern Africa*, ed. G. Kowero, B. M. Campbell, and U. R. Sumaila, 401–422. Jakarta: Center for International Forestry Research.

Thangata, P. H., and P. E. Hildebrand. 2012. Carbon Stock and Sequestration Potential of Agroforestry Systems in Smallholder Agroecosystems of Sub-Saharan Africa: Mechanisms for 'Reducing Emissions from Deforestation and Forest Degradation (REDD+). *Agriculture, Ecosystems and Environment* 158: 172–183.

Tobacco Industry Marketing Board. 2014. Annual Statistical Report. Accessed June 5 2016 http://www.timb.co.zw/downloads/2014%20Annual%20Statistical%20Report%20 final.pdf

Tobacco Industry Marketing Board. 2015. Annual Statistical Report. Accessed June 5 2016 http://www.timb.co.zw/downloads/2015%20Annual%20Statistical%20Report.pdf.

Tobacco Industry Marketing Board. 2018. Annual Statistical Report. Accessed October 17 2020 https://www.timb.co.zw/storage/app/media/Annual%20Stats%20 Report/2018%20ANNUAL%20STATISTICAL%20REPORT.pdf

Whitlow, J. R. 1980. Deforestation in Zimbabwe: Some Problems and Prospects. Accessed August 18 2017 https://opendocs.ids.ac.uk/opendocs/handle/123456789/9911

Woods, M. 2005. *Rural Geography*. London: Sage Publications, Ltd.

Part II

Demographic Diversity

6 The Changing Rural Periphery
Contested Landscape, Agricultural Preservation, and New Rural Residents in Dakota County, Minnesota, U.S.A.

Holly Barcus and David A. Lanegran

Introduction and Background

> A strong argument can be made that no issue is more important than the future of U.S. agriculture – and thus to our food supply and social sustainability – than identifying who will be the next generation of farmers who will steward the land and produce our food.
>
> (Hamilton 2010, 523)

Just beyond the edges of the urban metropolitan complex lies a zone of transition, characterized by large farming operations with big red barns and new white silos, acres of corn, soy or wheat, and the occasional suburban development, complete with smaller residential lots, very large homes, and grand entrance gates with bucolic names to appeal to potential buyers. This landscape is not uncommon in the Rural–Urban Interface (RUI) areas of the U.S. Midwestern states as urban areas expand spatially, periphery land values increase, and long-term farming operators sell out as younger generations move away from farming.

Although these changes are increasingly visible on the landscape, the RUI has long been a zone of transition, one that has often been overlooked not least "…because simplistic notions of sprawl reify and obscure, rather than illuminate, the complexity of economic and social spatial forces shaping the edge" (Audirac 1999, 7). Others focus specifically on the RUI farmland conversion processes in this transition zone (see Ilbery 1985; Bryant and Johnson 1992). More recently, however, this zone has become increasingly interconnected, and this blurs the spatial and social boundaries between urban and rural places (Lichter and Brown 2014; Lichter and Ziliak 2017). "The 'new' rural America is marked by accelerated spatial interdependence – a rapid blurring of traditional rural-urban spatial and symbolic boundaries" (Lichter and Brown 2014, 1) and is highly differentiated with regard to ethnoracial diversity (Lee and Sharp 2017), economies, housing types, and varied urban–suburban–rural identities (Garner 2017).

In this chapter, we evaluate the RUI in Dakota County, Minnesota, drawing a critical eye to the demographic and agricultural forces shaping the evolution of land uses in this transition zone. Our work is situated around a central question: In the context of changing demographic diversity in rural America and in the RUI specifically, how do agricultural land protection policies facilitate

DOI: 10.4324/9781003110095-8

the movement of ethnoracial minorities and other "new" farmers into farming practice? We argue that three threads of change affect new agricultural initiatives, including demographic change, land policies, and demand for local foods. We evaluate a case study of Dakota County, Minnesota, reflecting briefly on two examples to highlight these linkages.

The remainder of this chapter is organized first around the conceptual and theoretical underpinnings of the RUI changes occurring in the United States. The next section overviews our methods and approach, followed by our case study. Discussion and Conclusion sections follow the case study.

Conceptual Framework

Conceptually, we draw on several literatures to inform our understanding of the drivers and outcomes of land change in Dakota County. First is the RUI literature, which highlights the increasing connectivity and interdependencies between urban and rural places, rejecting the notion of the urban–rural binary in favor of a more nuanced understanding of these complex changes including social, economic, agricultural, symbolic, or perceptual changes. Second, we look to agricultural changes in the rural periphery and new demands by urban populations for local foods. For example, Inwood and Sharp (2012) following many others, highlight the increasing diversity of farm types in the RUI areas – noting that adaptation in the RUI is linked to external pressures, such as globalization or changing local food markets, as well as to household-level decisions about farm succession. Last, we look at changing rural demographics. This includes the aging of rural populations, and farm operator populations specifically, the increasing ethnoracial and cultural diversity of the RUI, and the new demographic characteristics of commuter populations.

The Contested RUI

These themes reveal interwoven threads of change occurring in the U.S. RUI. As Taylor notes "[e]xurbia captures the phenomenon of very-low-density, amenity-seeking, post-productivist residential settlement in rural areas" (Taylor 2011, 324). Such permanent settlements necessitate changes in land use regulations, which often also require the shifting of once open space or agricultural areas to residential developments. These shifts are characterized by both political and local complicity for change, such as new policies that allow land to be sold and converted from one use to another, as well as changing perceptions and preferences of both new and long-term residents in regard to land use. Such land use policies are decided upon generally at local scales, albeit with legal and political frames that may be orchestrated by larger, more spatially extensive governance frameworks, such as county or state-level policies. The legal and policy framework thus forms one aspect of land use policy in the RUI. A second aspect includes the changing desires of residents. Long-term residents may have one vision or narrative of the land, landscape, and livelihoods that are acceptable in the RUI, while new residents may hold vastly different perceptions and attitudes toward "appropriate" and "desirable" land uses. It is at the intersection of policies for change (or stasis) and diverging imaginations of desirable land uses where significant contest and conflict emerge (see Furuseth and Lapping 1999; Gosnell and Abrams 2011; Taylor 2011; Hiner 2016). As Furuseth and Lapping (1999, 3) note,

> Among the most contentious and long standing North American countryside planning issues is agricultural land alienation. Over the past 30 years, local, state, provincial, and national governments have examined and taken myriad actions to slow the loss of critical agricultural resources to urbanization. As we move into the next millennium, the challenge of farmland conversion remains a salient land use issue, perhaps the most common and use policy issue throughout Canada and the United States.

Farmland Preservation

Emerging with the growing environmental movement in the 1970s, concern over farmland preservation, particularly in rapidly urbanizing places gained sustained interest from environmentalists and urban planners (Pfeffer and Lapping 1994). Through the 1980s and the 1990s, proposals to stem the conversion of arable farmland near urban areas included greenbelts, urban growth buffers, Transfer of Development Rights (TDR), and Purchase of Development Rights (PDR) as tools for farmland and open space preservation. TDR, in particular, has received a fair bit of academic attention, although several authors concede that, while TDR can be effective it is dependent upon the location and the mix of specific policies incorporated within the TDR protocols (see Kaplowitz et al. 2008; Pruetz and Standridge 2008). Other open space preservation strategies include conservation easements and land trusts (see Daniels and Lapping 2005). Local land policies are thus one strategy for preserving agricultural and open land in the RUI.

At the farm level, there are additional challenges. Rising land values and declining interest in farming within existing farm families can make selling the family farmland attractive, making farm succession decisions a challenge to the maintenance or preservation of farmland in the RUI (Inwood and Sharp 2012). Additionally, the RUI farm types are diverse, including traditional farms as well as those adapted to respond to urban-oriented markets, such as direct sale, organic, speciality crops, and nursery/greenhouse operations (Jackson-Smith and Sharp 2008). Some farms are able to leverage agritourism businesses to create new products and markets that complement the existing farming operations, but other farms face growing disdain of new suburban neighbors over farm smells or noise.

This diversity reflects not only changing land use pressures but changing demands for food. "In 2002, 55 percent of all farm sales were from farms located in counties at the rural-urban interface, even though 60 percent of all farmland is located outside RUI counties" (Jackson-Smith and Sharp 2008, 3), and the growth of the number and sizes of farmers markets across urban and suburban areas underscore continued demand for these local foods. Agricultural areas in the RUI thus seem essential areas for continued food production even though the farm-to-table or farm-to-farmers market modes are transforming the scale and transport of food.

Demographic Change

Beginning in the early 2000s it became clear that rural areas were continuing to see significant demographic change, with cultural, racial, and ethnic diversity added to long-standing trends of out-migration, aging, and in some places, selective amenity-based in-migration. Beyond the warnings that rural places were experiencing population decline, out-migration, and aging, was a notable countertrend of migration to particular types of rural areas, such as amenity-rich areas and similar retirement-type destinations (Johnson 2012). Rural areas are also becoming increasingly diverse racially and ethnically, with ethnoracial minorities accounting for 83% of rural growth between 2000 and 2010 (Johnson 2012). As Barcus and Simmons (2013) note, the trajectories of migrants from different ethnoracial backgrounds, foreign- and domestic-born, are highly differentiated and include demographic groups at all levels of the economic stratum, education, and occupational backgrounds. The RUI also attracts a diverse population depending on factors such as housing, proximity to urban amenities, location-specific rural amenities, and land uses. Exurban areas, for example, include long-distance commuters, while suburban communities may be attractive for their schools and commercial centers. "The concept of exurbia has traditionally been used to describe settlement patterns simultaneously dispersed from the city yet also connected to urban networks" (Taylor 2011, 323) "… and are characterized by an influx of residents seeking specific idealized notions of rural life…." (Hiner 2016, 525). Further, the scholarly focus on farmer aging and the process of succession and adaptation, and specifically decisions made by individual families as to the future of their farm operation, has implications for land use, particularly in the RUI (Inwood and Sharp 2012).

Taken together, these threads, increasing demographic diversity and aging of existing farmers, protection of farmlands and a desire to continue to maintain rural landscapes near cities, and new and more diverse imaginations of the purpose of the RUI, result in the increasingly contested RUI geographies.

Methods and Approach

In this chapter, we take a political ecology approach to understanding the multiple threads of change co-occurring in Dakota County, Minnesota (see Figure 6.2 for location). Broadly, a political ecology framework helps to conceptualize the varying scales and actors that influence changing land practices, and a case study approach seeks to elucidate micro-scale outcomes for a particular place. Robbins suggests that

> …political ecologists follow a mode of explanation that evaluates the influence of variables acting at a number of scales, each nested within another, with local decisions influenced by regional policies, which are in turn directed by global politics and economics.
>
> (Robbins 2012, 20)

We develop our case study through this multi-scalar lens, acknowledging that a case study is limited to a specific place. However, given the complexity of change in the RUI, selecting a singular place to begin teasing out the inter-related factors that are influencing the emergence of new farmers seems appropriate. Case studies, as described by Yin (2009, 2), typically answer "how" or "why" questions about contemporary phenomena, "…an essential tactic is to use multiple sources of evidence with data needing to converge in a triangulating fashion".

In keeping with this multi-scalar case study approach, we began our research with a review of Dakota County historical documents and land use plans and follow-up interviews with key experts and local leaders, including land conservation experts, political representatives, and knowledgeable local activists and educators.

We also utilized existing statistical data resources such as the U.S. Decennial Census (2000, 2010) and American Community Survey (U.S. Bureau of Census 2017) and the U.S. Agricultural Census (2002, 2007, 2012, 2017). While these data are easily mappable with GIS, they are limited by their lack of geographic specificity as it pertains to ethnoracial groups in the RUI. As these groups are small, the total numbers of different groups may be suppressed in the data. To better assess new populations on the landscape, we utilized an additional data source, the Association of Religious Data Archives (The ARDA 2020), which allows congregations of different faiths to be mapped. These data come from the 2010 Religious Congregations and Membership Study conducted by the Association of Statisticians of American Religious Bodies (The ARDA 2020). Since the U.S. Government does not collect statistics on religious affiliation or memberships, this dataset represents one of the few datasets about the U.S. religious groups (The ARDA 2020). The ARDA thus provides a snapshot of emerging populations that might not be fully identified in the U.S. Census.

To better understand new farmers on the landscape, we chose to interview representatives from two new farming operations. These operations are examples of new populations emerging as farmers in Dakota County. There are, however, additional Community-Supported Agriculture (CSAs) and niche farmers on the landscape both in Dakota County and spread across the seven-county metropolitan area of Saint Paul-Minneapolis, Minnesota, and these farmers are not necessarily from ethnoracial minority groups, although they may be new farmers.

Last, we utilize landscape observations to more fully develop our understanding of change in this area of the RUI. Specifically, the co-authors have individually and together driven through various transects of Dakota County, watching as the landscape evolves and documenting some of these changes with photographs.

Case Study – Dakota County, Minnesota

The Saint Paul-Minneapolis Minnesota metropolitan area is characterized by population growth and expansion of its urban populations. The urban area comprises seven counties, inclusive of the urban core, and is colloquially known as the Twin Cities. Dakota County, one of the seven metropolitan counties, specifically saw a significant population increase between 1960–1970 and again

Table 6.1 Regional Population Growth, 1900–2017

Year	Dakota County pop	% Change Dakota County[a]	Seven-county metro area	% Change metro area**	Minnesota state population	% Change MN[a]
1900[1]	21,733				1,751,394	
1920[1]	28,967	33.3			2,387,125	36.3
1940[1]	39,660	36.9			2,792,300	17.0
1950[1]	49,019	23.6			2,982,483	6.8
1960[1]	78,303	59.7			3,413,864	14.5
1970[1]	139,808	78.5	1,874,612		3,804,971	11.5
1980[1]	194,279	39.0	1,985,873	5.9	4,075,970	7.1
1990[1]	275,227	41.7	2,288,721	15.3	4,375,099	7.3
2000[2]	355,904*	29.3	2,288,721	15.4	4,919,479	12.4
2010[2]	398,552*	12.0	2,849,567	7.9	5,303,925	7.8
2017[2] (est)	422,580*	6.0	3,075,563	7.9	5,576,606	5.1

Source: [1] U.S. Census Bureau (1995). [2] 2000–2017: U.S. Bureau of Census (2000, 2010, 2017).
a Percentage Change calculated based on preceding decade listed in the table (% change 1950–1960 listed in 1960 row).
** Data Source = U.S Bureau of Census Decennial Census compiled by Metropolitan Council (Met Council 2020) https://stats.metc.state.mn.us/profile/detail.aspx?c=037#POPANDHH

1970–1980 (Table 6.1), with growth continuing until the present time period. Most of the initial growth in Dakota County, however, occurred in near proximity to the Twin Cities, at the northern edge of the county, as transportation systems developed and commuting became more feasible. The spatial extent of this growth has rapidly expanded to the central and traditionally agricultural regions of the county. In response to increasing populations and urban expansion in the late 1990s, the county initiated a process of agricultural and natural area protection which has facilitated a new imagination of agriculture and of agricultural entrepreneurs as well as farmland preservation.

At the state level, increased population growth (Table 6.1) is accompanied by increasing diversity (Figure 6.1). The top five cultural groups following White, non-Hispanic in Minnesota include African Americans, Mexican, Native American, Hmong, Somali, and Indian, all exceeding 50,000 in population (Minnesota Compass A 2020). Although the Latinx population is one of the largest minority populations in Minnesota, it is notable that the growth of this population is dispersed across the state (Figure 6.2). When taken together, these maps confirm that diversity in Minnesota includes a wide range of ethnoracial minority populations.

Demographic change is closely linked to evolving agricultural practices and markets in the Twin Cities, which has witnessed an explosion of demand for local foods. For example, Minnesota Grown currently lists 82 CSA farms in Minnesota and 97 farmers markets within 50 miles of Saint Paul (Minnesota Grown 2020). An important factor in the success of local farmers' markets around the

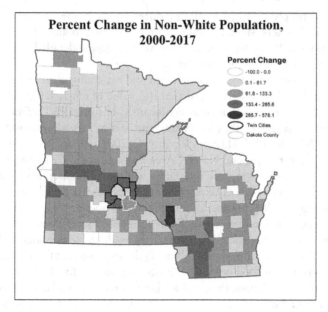

Figure 6.1 Percent Change in Non-White Population, 2000–2017.
Source: U.S. Bureau of Census (2000, 2010, 2017).

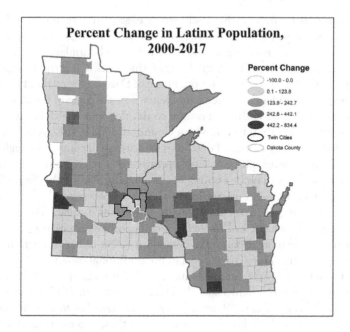

Figure 6.2 Percent Change in Latinx Population, 2000–2017.
Source: U.S. Bureau of Census (2000, 2010, 2017).

metro area is the provision of locally produced foods. Some markets, such as the Saint Paul Farmers Market, require all foods to be produced by the seller, and applications to become a vendor at the market require detailed maps of and directions to growing fields for verification.

The combination of increasing population and expanding developments and greater demand for local foods means that the value of fertile farmland within proximity to the metro area has risen significantly. Historically, Dakota County was agricultural. In the 1950s, 85% of the land was in agricultural production, declining to 60% in 1997 (Dakota County C 2002, 9). In 2002, when the "Dakota County Farmland and Natural Area Protection Plan" (Dakota County C 2002) was adopted, there were 890 farms in Dakota County, averaging 249 acres (100.8 hectares) each (Dakota County C 2002, 9). Development pressure increases land values and, as long-term farmers and farming families begin to leave farming, the land previously farmed potentially becomes available for development. Increasing demand by Twin Cities residents, restaurants, and other businesses for local foods, however, is allowing the expansion of farmers markets, CSAs, and other local food producers. Since the market for these products is centered on the Twin Cities, the need for land to remain available for existing and expanding niche farming operations is an important factor for these food and farm product businesses.

Contextualizing Demographic Change in Dakota County

Dakota County is located on the southeastern edge of the Twin Cities. In the early 1990s, the County embarked on a new land use planning process (discussed in more detail below) resulting in the aforementioned Farmland and Natural Areas Program (FNAP) plan. At the heart of this process was the recognition that significantly different visions for the future of Dakota County land use were held by long-term residents and newer residents. The outcome of the land planning process recognized both the pressure to develop and the desire to retain open space and agricultural land (Dakota County C 2002). This process and its resulting plan have had a range of outcomes that are significant to the region.

The changing demographics of farm operators is a micro-trend within the broader increase in small niche farming operations surrounding the Twin Cities. Dakota County becomes an interesting case study in observing the interlinked changes co-occurring in these traditionally farmed RUI areas. Three demographic changes are salient in Dakota County. First, most population growth is occurring not in the more distant rural communities but rather through increasing commuting to the Twin Cities. Second, the traditional owners of the land in this region are farmers, who are predominantly White and aging, making farm succession or transition salient issues. In Dakota County in 2017, for example, there were 820 farms on 227,081 acres (91,896.4 hectares) of land with a corresponding 1,338 individuals listed as producers. Of these producers, 29.7% were aged over 65 years (61.6% over age 55) (Census of Agriculture 2017). Continued aging of the producer population means that land will either be transferred to a new generation or may become available for sale or rezoning.

Last, as suggested by the emergence of new religious affiliations, Dakota County is increasingly culturally diverse. As indicated in Table 6.2, we turn to alternative sources of data that reflect changes in local populations. Between 1990–2000 and again 2000–2010, the growth in new religious institutions and religious adherents grew significantly in rural Dakota County, indicating the presence of new populations. While this isn't a direct measure of population size, the presence of these new, and sometimes quite large, religious institutions (Figure 6.3) on the landscape are indicative of the presence of a more diverse population.

The demographic changes taking place in Dakota County, however, are certainly not sufficient conditions to produce changes in land opportunities for new or niche farmers in the RUI. Nor is it the case that all new populations have farming interests. Rather, these factors, in combination with innovative and

Table 6.2 New Rural Populations in Dakota County as Represented through Changing Representations of Religious Entities

	1980	1990	2000	2010
Buddhist	0	0	3 congregations[a]	2 congregations[a] 696 adherents[a]
Baha'i	0	0	84 adherents[a]	114 adherents[a]
Hindu	0	0	0	2 temples[a] 190 adherents[a]
Muslim	0	0	0	5 congregations[a] 1,540 adherents[a]

Source: The ARDA (2020).
a Terminology reflects survey data collection frames.

Figure 6.3 Buddhist Temple in Dakota County, Minnesota.
Source: Authors.

progressive land use policies have opened opportunities for new imaginations of farming in this zone. In the next section, we lay out the history and context in which Dakota County has re-envisioned its land use policies.

Farmland Protection and Comprehensive Land Protection Plan

During the past two decades, the Dakota County government implemented an innovative land conservation plan, developed in response to residents' concern about changes in the landscape (Dakota County C 2002). Urbanization brought dramatic changes to county land use patterns during the 1990s. By the end of the decade, only 2%–3% of the original pre-settlement natural areas remained in the county (Dakota County C 2002). The remaining areas had very uncertain futures.

In 1999, the County staff with a $200,000 grant from the State Legislature began a three-year effort to develop a plan to preserve natural areas and farmland. Part of the planning process consisted of a survey of county residents' attitudes about land protection. The results were astounding. Ninety-six percent of respondents favored protection of natural areas and lakes, while 56% supported the preservation of agricultural land (Dakota County C 2002). This strong support made it possible for the County Board to adopt The Farmland and Natural Area Protection Plan in January 2002. Now titled the Land Conservation Program, the plan identified 78,000 acres (31,565.5 hectares) of land for protection, almost equally divided between farmland and natural areas. The preservation plan involves the paying of landowners who enroll their property in the program and agree to maintain its natural or agricultural character.

A funding source had to be identified to implement the preservation plan. County leadership turned to the residents for guidance by conducting a survey to see if the taxpayers would support a referendum enabling the county to borrow money to implement the plan. The survey showed strong support and the subsequent referendum to authorize the County Board to levy a tax to support the plan passed by a sizable majority (Dakota County C 2002). It is not possible to know why individuals voted to support this additional tax, but it is likely that residents acted out of concern for water quality in the county as well as a more general appreciation for the natural landscape. Once the plan was adopted and county tax revenue was authorized it was possible for the County to acquire additional financial support from landowner donations, other units of government, and philanthropic organizations (Dakota County C 2002).

There are four components of the comprehensive plan to promote Land Conservation. The Land Conservation Agricultural Land Stewardship Program acquires fee titles or permanent easements from willing sellers on lands that include agricultural use to promote water quality and wildlife habitat. The Land Conservation Natural Area Protection Program acquires fee title and/or easements from willing landowners with required natural resource management plans to permanently protect, connect, and enhance natural areas of state, regional, or County significance. The staff of the county ensures that the Stewardship Plans for specific areas are maintained.

Land conservation in Dakota County is also supported by important initiatives by other levels of government, especially the Metropolitan Council. Planning in the Twin Cities Metropolitan Area is unique. This is in large part due to the adoption of the Metropolitan Land Planning Act by the state of Minnesota and the subsequent formation of the Metropolitan Council in 1967 (Met Council 2020a). The Council's function is to ensure the orderly and economic development within the seven-county metropolitan area. The current regional development plan, called Thrive MSP 2040, established a regional vision and adopts land use development policies through 2040 (Met Council 2020b). The comprehensive plan reflects regional policies at the same time as identifying important local goals and objectives. The plan includes six categories of land use, three urban and three rural. One category, "permanently agriculture", is designed to preserve large swaths of farmland areas with prime agricultural soils (Dakota County C 2002). About half a million acres (202,342.8 hectares) are planned and zoned by local communities to maintain agriculture as the primary long-term land use (Met Council 2020b). The future of this agricultural area depends on cooperation among several levels of government. The regional planning authority's role is to promote the use of the Agricultural Preserves and Green Acres programs by supporting local efforts that maintain agricultural land uses through 2040. The agency will also promote agricultural practices that protect the region's water resources and provide economic opportunities for farmers and promote local food production. Under this partnership, the county and township governments must develop and implement strategies for protecting farmlands, such as exclusive agricultural zoning, agricultural security districts, and lower residential densities. They are also asked to consider opportunities for smaller-acreage agricultural operations to support food production for local markets (Met Council 2020b).

At the state level, there are two Minnesota taxation policies designed to help preserve agriculture in the metro area. The Metropolitan Agricultural Preserve Program was established by the Minnesota Legislature in 1980 (see Dakota County A (2020) for more program details). Property zoned long-term agricultural by the local community, with a maximum residential density of one house per 40 acres (16.2 hectares), is eligible for this property tax program. Owners sign an eight-year perpetual covenant/agreement to leave the property in agricultural use and farm using acceptable practices as approved by the County Agricultural Service. In return, owners pay annual property taxes based on the agricultural market value only. The market value of land in the Agricultural Preserve is based on sales of agricultural property in non-metropolitan counties as determined by the Minnesota Department of Revenue.

Dakota County has also helped maintain agricultural land use with a property tax deferral program, called Green Acres (Dakota County C 2002). Agricultural property devoted to the "production for sale of agricultural products", may be eligible for the Green Acres program if it meets three criteria. It is at least 10 acres (4.0 hectares) in size, is primarily devoted to agricultural use, must be occupied by the farmer, or the property has to have been in the applicant's family for at least seven years. The County Tax Assessor determines two values on Green

Acres property. The "actual market value" is based on sales of similar property, which may be influenced by urban development pressures. The "agricultural value" is based on sales of agricultural property not impacted by other influences, such as development. Taxes are calculated on both market values but paid on the lower, agricultural value each year. The difference between the tax calculated on agricultural market value and the actual market value is deferred until the property is sold or no longer qualifies for the program. Farmland preservation in Dakota County is the result of a combination of planning controls, subsidies of landowners, and the public acquiring permanent easement or direct ownership of parcels. All of which fit into the Comprehensive Plan.

New Farm Operators and Agricultural Entrepreneurs

In this chapter, we are exploring the combined processes of land change and demographic change and the potential implications this might have for opportunities for new agricultural entrepreneurs. The combination of population growth and expansion outward from the metropolitan core and broadscale changes in demographic diversity occurring across the state of Minnesota suggests that we should expect local and regional business, including agriculture-based entrepreneurs to become more diverse as well. In this section, we explore these trends utilizing data from the U. S. Department of Agriculture. As Figure 6.2 (above) illustrates, Minnesota's increasing demographic diversity is widespread and occurs at all levels in the urban–rural hierarchy. We also see the percentages of non-White, principal farm operators begin to grow slightly between 2002 and 2012, although the percentage of the Latinx Principal Operators fluctuates (Table 6.3).

Utilizing data from the U.S. Census of Agriculture, we also mapped the Simpson Diversity Index values to assess diversity among farm operators (Figure 6.4). The Simpson Index measures the diversity within counties that is comparable to the index values of other counties in the map. Index values range from 0 to 1, 0 indicating no diversity and values greater than 0 indicating increasing diversity. The index ultimately indicates the probability that two randomly selected individuals will come from different population groups (Plane and Rogerson 1994). We chose to use principal farm operators as an indicator of managerial expertise and influence in an operation, rather than simply mapping workers in the industry. We felt that mapping operators reveal a level of entrepreneurship and opportunity that might not be represented by simply mapping employees of agricultural firms. In Minnesota, some of the highest levels of diversity in farm operators are found in the seven-county metro area. Dakota County has a score of 0.11, indicating a moderate level of diversity among farm operators, as compared to the other diversity values present in Minnesota and Wisconsin in 2012. In 2012, Dakota County had 896 farms, of which 842 (93.9%) were operated by White operators, 47 (5.2%) by Asian operators, three (0.33%) by American Indian operators, and four (0.45%) Hispanic operators. Clearly, the overall diversity is relatively low. However, by 2017, the number of farms had decreased to 820 farms covering 227,081 acres (91,896.4 hectares). Diversity, however, had

Table 6.3 Percent Change in Farm Operators, 2002–2017

	Total farm operators, 2002	% Non-White farm operators, 2002 (% HISP)	Total farm operators, 2007	% Non-White farm operators, 2007 (% HISP)	Total farm operators, 2012	% Non-White farm operators, 2012 (% HISP)	Total farm producers, 2017[a]	% Non-White farm producers, 2017 (% HISP)
Minnesota	80,839	0.3% (0.62%)	80,992	0.7% (0.37%)	74,542	0.75% (0.45%)	111,760	0.84% (o.58%)
Wisconsin	77,131	0.4% (0.68%)	78,463	0.5% (0.31%)	69,754	0.5% (0.41%)	110,347	0.87% (0.59%)

Data Source: USDA (2002, 2007, 2012), "Select Principal Operator Characteristics by Race" and "Selected Operator Characteristics, Spanish, Hispanic or Latino Origin"; USDA (2017), "Selected Producer Characteristics" Table 52.
a 2002–2012 data reflect only Principal Operators, 2017 includes all operators.

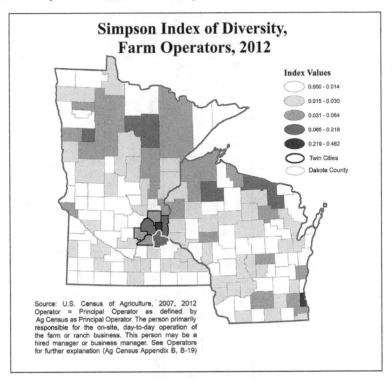

Figure 6.4 Simpson Index of Diversity of Farm Operators, 2012.

Source: U.S. Census of Agriculture (2007, 2012).

increased. In 2017, of the 1,349 producers,[1] 11 (0.8%) producers identified as Hispanic, two (0.15%) as American Indian, 94 (7%) as Asian, 1,240 (91.9%) as White, and two (0.15%) as multi-race. Importantly, however, of the 56 farms owned by Asian producers, the average size of the farm was 18 acres (7.3 hectares), while for White farmers, owning 764 farms, the average size was 296 acres (119.8 hectares) (U.S. Census of Agriculture 2017).

We use the following two examples of new farm projects to illustrate how the co-occurrence of demographic change, progressive land policy, and local food demands have opened opportunities for new farmers and specifically ethnoracial minority farmers in Dakota County. So far, we have demonstrated state and metropolitan-level demographic change and growth and discussed state and county-level programs of land regulation regarding agricultural land preservation. But what does this look like in the landscape at the micro-scale? The Hmong American Farmers Association (HAFA) and the Main Street Project are two examples of outcomes of these processes. We offer only brief descriptions to illustrate the multi-scalar nature of change and argue for continued interrogation of the linked policies and trends that facilitate new farm models and opportunities for new farmers.

The HAFA Farms

The first example is the HAFA, established in 2011 by a group of Hmong American farming families (HAFA 2019). As part of a joint project to further develop the farming infrastructure, support and training of Hmong farmers, the HAFA had a silent partner interested in fostering a new agricultural model with a social justice angle. The partnership allowed the HAFA to establish a 155-acres (62.7 hectares) research and incubator farm located in Vermillion Township, Dakota County through a lease agreement as well as collaboration with the County. The land is divided amongst families who individually farm their own small plot of land and collaboratively contribute to the HAFA CSA as well as having options for home consumption and sales at local farmers markets. The HAFA helps farmers access bi-cultural and bi-lingual trainings to improve land stewardship and promote organic farming methods. Their products include CSA shares, farmer market sales, and home consumption.

Hmong farmers have a long history in Minnesota, beginning with the arrival of the first Hmong refugees in 1976 (Vang 2008, 12), resettled largely from refugee camps in Vietnam following the Vietnam War. The population grew quickly, through both additional refugee resettlements as well as internal secondary migration to Minnesota from other resettlement destinations. The population has been socially and economically mobile and has dispersed across both the urban and suburban communities in the metro area (Vang 2008). By 2000, nearly 54% of the Hmong population in Minnesota owned their own homes (Vang 2008, 16) – an important measure of social mobility in the United States. Numbering nearly 81,966 (2013–2017) in total with 95.4% residing within the Twin Cities (Minnesota Compass 2020), this population is growing numerically and is increasingly affluent and influential within the region.

Hmong farms are a key contributor to a growing local food movement as long-term producers of local foods, both for home consumption and local sale at ethnic and farmers markets. Early Hmong immigrants came from a largely agricultural background in Laos. In the early 1980s, the Office of Refugee Resettlement identified three categories of farm initiatives oriented toward moving Hmong toward self-sufficiency. As summarized by Vang (2008), these included community gardening for home consumption, and commercial sales to supplement household incomes and farming at a commercial level (Downing et al. 1984). State-level programs, administered by a range of agencies such as the University of Minnesota Extension Service, were also created, including the Minnesota Agricultural Enterprise for New Americans (MAENA) and the Hiawatha Valley Farm Cooperative (HVFC). These initial projects had mixed results (Tsu 2017).

Niche farming by and for this population is a reflection of both entrepreneurialism and cultural continuance while also providing degrees of food security and economic security and contributing to local food provision in the Twin Cities. However, despite their success, one challenge for any new niche farmer wanting to farm within close proximity to the urban area is the cost of land. Ethnic minority and immigrant farmers lack historic ties to the farming communities and networks and quite often are not prepared to purchase extensive land acreages, especially given the higher land costs in the RUI.

The Mainstreet Project

The second example is the Main Street Project, located in Northfield, in Dakota County, recently renamed Regeneration Farms to better reflect the innovative and evolving nature of their farm model (see Casillas 2019). The momentum of the ideas behind Regeneration Farms has sparked global interest, but here we focus on the initial site of the farming incubator, the Main Street Project. The Main Street Project promotes a poultry-centered farming model nick-named "tree range chickens". In this project, chickens coexist with a range of other on-farm activities referred to as the farm ecosystem. The project promotes the idea that chickens are an affordable and adaptable element in a farm system which includes rotating a variety of crops on half-acre production units. The model is intended to be accessible to limited resource farmers, and specifically to Latinx and immigrant farmers. The project has recently expanded to include a new 100-acre demonstration project for training, research, and development of farming model.

As is clear from Map 2, the Latinx population is expanding rapidly and is one of the fastest growing in Minnesota, increasing from 54,000 in 1990 to 271,000 in 2014 (Hartzler 2014). The population is clustered in the southern portion of the state, attracted initially by employment in large-scale agricultural farming processing, and manufacturing plants. The majority are native born and despite increasing high school graduation rates, nearly one in four live in poverty (Hartzler 2014). Latinx farmers have limited access to land so cooperatives such as the Main Street Project or Agua Gorda in Long Prairie Minnesota help fill this gap and provide training (Meersman 2014), thus facilitating opportunities for new farmers from this population.

Both the HAFA and the Main Street Project exemplify farming models that are community-driven and seek to empower individuals with farming backgrounds who lack access to farmland, or capital. Both, to different degrees, also seek to foster a farm model that is environmentally and socially sustainable. The HAFA operates as a cooperative for farm families who individually plant and manage their plot of land but work collaboratively for marketing and training opportunities. The Main Street Project focuses on a single model that all participants adopt but which is also supported by research and training for the farmers.

Discussion

In this chapter, we outline three co-occurring factors that are influencing land use transitions, specifically agricultural transitions in the Twin Cities RUI. Land use policies in the United States are highly differentiated by scale – there are federal (i.e. U.S. Farm Bill), state, regional, county, and sometimes city or township levels to land use policies and processes. Dakota County is no exception. In Dakota County, we see the influence of state, metropolitan, and county-level land use ideals and policies. We argue, however, that Dakota County represents an interesting case study in how the RUI in a rapidly growing metropolitan area can encourage agricultural land preservation while simultaneously encouraging the development of new forms of farming – forms that are more accessible to farmers interested in niche farming at a smaller scale. By creating land use policies that allow new expressions of agriculture and new forms of leasing and ownership, the RUI in Dakota County is maintaining a rural feel, preserving agricultural land, and opening opportunities for the options of continuance, departure, or succession of traditional large acreage commodity cropland to smaller niche farming operations.

Social-Economic and Environmental Impacts

By definition, land use policies in the RUI seek to serve a highly diverse group of residents, including those who occupy more rural or open areas as well as those desiring a more suburban, developed landscape. One of the key components of Dakota County's land use planning process in the late 1990s was widespread support for the protection of agricultural land and natural areas – preservation that presented itself as being as much about the protection of landscapes as about land uses. Concern over the changing landscape was an important driver of the initiation of the land conservation plan in the 1990s, as this was a time when 2,000–3,000 acres (809.4–1,214.1 hectares) per year were being converted from primarily agricultural to suburban development (Dakota County C 2002). One county-level expert suggested that common ground across a range of resident types was the importance of preserving water quality. Water quality protection, when combined with natural area and agricultural protection, has resulted in the establishment of extensive connected green areas in the county. These protected areas not only protect the landscape value of open or agricultural fields, they also provide corridors for wildlife and buffers around water bodies that offer protection from runoff and maintain habitat (Dakota County C 2002). In this case, the socio-economic and environmental aspects of land conservation are deeply intertwined in perceptions of landscape preservation.

Connection of Local to Global Scale

Although this chapter evaluates the implications of a small case study area, there are several broader lessons. First, in the case of Dakota County, our key informants noted that land use decisions could be controversial at local levels. Finding a common goal, such as water quality protection in this case, helped establish the means to imagine land use plans that met a range of social and environmental goals. As a voter-approved plan with financial support from voter-approved property tax increases, combined with matching funds from the Minnesota Legislature, the plan had local and state-level support (Dakota County B 2020). Thus, the combination of a successful collaborative public planning process and common set of goals has facilitated the implementation and success of the FNAP.

One of the outcomes of the plan, as described here, is a different imagination of how farmland could be used and, importantly, of how to continue to imagine and reimagine exurban areas. Through collaborative work between property owners, the county, and potential new farmers, Dakota County has been able to help facilitate plans that meet the County's desire for protecting water quality and the different imaginations of new farmer arrangements. The Dakota County FNAP has won several notable awards, including from the American Planning Association, Minnesota Association of Counties, Minnesota Environmental Initiatives and the 2009 Governor's Award for Pollution Prevention (Program History 2017) and the plan itself is notable for its community engagement process and collaboration. Agricultural land plays an important role in the landscape and natural processes of the county. Exurbia and the RUI is a global phenomenon (Woods 2007; Taylor 2011). Creating pathways for new farmers, as well as options for farm transitions, will continue to be a pressing issue in the RUI.

Changing Ethnoracial and Cultural Diversity in the RUI

Dakota County provides a compelling case study of the intersecting changes occurring in the RUI in America. While this chapter focuses specifically on land policy changes and diversity amongst new agricultural entrepreneurs, it is important to note the trends evident in Figures 6.1 and 6.2 that reflect the increasing ethnoracial and cultural diversity that is occurring in the RUI of the Twin Cities metro area, regardless of agricultural change. Numerous threads of change are captured in these dynamic areas surrounding the metropolitan core. For example, ethnoracial and cultural diversity has roots in a range of other co-occurring processes in this area. The Twin Cities is home to a wide range of higher education institutions, which attract diverse students and faculty, and corporations that also foster greater international diversity. Refugee populations, such as Hmong and Somali populations, among others, further add to growing diversity in the metro area. Each of these different population groups arrives with or aspires to greater social and economic mobility, and the RUI areas may offer housing and education opportunities that fit these aspirations. The U. S. population as a whole is becoming significantly more diverse, with urban areas experiencing greater levels of diversity, much of which is fueled by natural increase, rather than by immigration alone (Mather and Lee 2020). As such, it is not surprising that the RUI too will be included in these changes. As we've tried to describe in this chapter, the confluence of increasing diversity broadly, combined with demand for local foods and progressive land use policies are increasing access to farmland for new farmers in Dakota County, Minnesota. As the quote at the beginning of the chapter highlights, finding opportunities to support a new generation of farmers is essential to continue agricultural innovation and productivity. The Dakota County case study provides one example of how this is occurring in the United States.

Conclusions

In this chapter, we sought to link broadscale processes of change occurring in rural and the RUI areas to a micro-scale example of innovative farmland management in Dakota County, Minnesota. We illustrate how the confluence of broadscale factors affecting many rural U.S. communities, including demographic change (aging and diversifying populations) and farm transitions, coupled with progressive land use policies, and a growing demand for local foods, have opened opportunities for these demonstration type projects to foster training and land access for new, ethnoracial minority farmers in the RUI.

Land policies at the state and metropolitan level have been coupled with local-scale policies and collaborative planning to help new farmers find and manage land that allows for new visions of farming while supporting the transition of larger-scale farms or lands to new imaginations of farming and simultaneously working toward broader county-level environmental goals such as water quality and habitat protection. While the examples provided in this chapter are small-scale, they reflect substantial work to maintain a range of landscape and land use goals in the RUI. Additional questions, however remain, such as how can local land policies facilitate movement of smaller-scale agriculture operations into formerly large-scale agricultural operation land, when land is transitioning? Can land policies that favor land preservation provide incentives for families to lease land, at least in part to smaller land users who may not otherwise have access to farmland in close proximity to retail outlets? And, as urban demand for local foods grows, will we see a shift to larger-scale alternative farm product producers?

These questions require a broader-scale and longer timeline to interrogate but we are hopeful that the examples provided here can help incentivize continued imagination of new farming practices, policies, and collaborations in the RUI.

Note

1 Operators. The term "operators" has been replaced with the term "producers". (See 2017 Census of Agriculture, Appendix B, page B-16).

References

American Community Survey. 2017. U.S. Bureau of Census.

Audirac, I. 1999. Unsettled Views about the Fringe: Rural-Urban or Urban-Rural Frontiers. In *Contested Countryside: The Rural Urban Fringe of North America*, ed. O. J. Furuseth and M. Lapping, 7–32. Brookfield: Ashgate.

Barcus, H. R., and L. Simmons. 2013. Ethnic Restructuring in Rural America: Migration and the Changing Faces of Rural Communities in the Great Plains. *Professional Geographer* 65 (1): 130–152.

Bryant, C. R., and Johnson, T. R. 1992. *Agriculture in the City's Countryside*. Toronto: University of Toronto Press.

Casillas, R. 2019. *Why are we renaming Main Street Project?* Blog Post 10/15/19. https://sharing-our-roots.org/blog/why-are-we-renaming-main-street-project/. Accessed May 28, 2020.

Daniels, T., and M. Lapping. 2005. Land Preservation: An Essential Ingredient in Smart Growth. *Journal of Planning Literature* 19 (3): 316–329.

Dakota County A. 2020. Agricultural Preserves. Property Tax Programs, Homesteads & Credits Dakota County. 5/7/2020. Accessed June 26, 2020. https://www.co.dakota.mn.us/HomeProperty/TaxPrograms/AgriculturalPreserves/Pages/default.aspx

Dakota County B. 2020. *Land Conservation Plan for Dakota County. Pre-Public Review Draft March 9, 2020*. Accessed June 26, 2020. Final Plan available at https://www.co.dakota.mn.us/Environment/LandConservation/Plan/Documents/LandConservationPlan.pdf

Dakota County C. 2002. *Dakota County Farmland and Natural Area Protection Plan*. Land Conservation. Dakota County. Accessed July 1, 2020. https://www.co.dakota.mn.us/Environment/LandConservation/History/Pages/default.aspx

Downing, B. T., D. P. Olney, S. R. Mason, and G. Hendricks. 1984. *The Hmong Resettlement Study*. Minneapolis: University of Minnesota.

Furuseth, O. J., and M. Lapping (Eds). 1999. *Contested Countryside: The Rural Urban Fringe in North America*. Aldershot: Ashgate.

Garner, B. 2017. "Perfectly Positioned": The Blurring of Urban, Suburban, and Rural Boundaries in a Southern Community. *Annals, AAPSS* 672: 46–63.

Gosnell, H., and J. Abrams. 2011. Amenity Migration, Exurbia, and Emerging Rural Landscapes. *GeoJournal* 76 (4): 303–322.

HAFA (Hmong American Farmers Association). 2019. Our Story. HAFA. Accessed June 26, 2020. https://www.hmongfarmers.com/story/

Hamilton, N. D. 2010. America's New Agrarians: Policy Opportunities and Legal Innovations to Support New Farmers. *Fordham Environmental Law Review* 22 (3): 523–562.

Hartzler, N. 2014. Minnesota's Hispanic Population: 5 Interesting Trends. Minnesota Compass. MNCompass.org. Accessed July 7, 2020. https://www.mncompass.org/trends/insights/2014-10-02-hispanic-population-trends

Hiner, C. C. 2016. Beyond the Edge and in Between: (Re)conceptualizing the Rural–Urban Interface as Meaning–Model–Metaphor. *The Professional Geographer* 68 (4): 520–532.

Inwood, S. M., and J. S. Sharp. 2012. Farm Persistence and Adaptation at the Rural-Urban Interface: Succession and Farm Adjustment. *Journal of Rural Studies* 28 (1): 107–117.

Ilbery, B. W. 1985. *Agricultural Geography: A Social and Economic Analysis*. Oxford: Oxford University Press.

Jackson-Smith, D., and J. Sharp. 2008. Farming in the Urban Shadow: Supporting Agriculture at the Rural-Urban Interface. *Rural Realities* 2 (4): 1–12.

Johnson, K. M. 2012. Rural Demographic Change in the New Century: Slower Growth, Increased Diversity. *Carsey Institute Issue Brief* 44: 1–12.

Kaplowitz, M. D., P. Machemer, and R. Pruetz. 2008. Planners' Experiences in Managing Growth Using Transferable Development Rights (TDR) in the United States. *Land Use Policy* 25 (3): 378–387.

Lee, B. A., and G. Sharp. 2017. Ethnoracial Diversity across the Rural-Urban Continuum. *Annals, AAPSS* 672: 26–45.

Lichter, D. T., and D. L. Brown. 2014. The New Rural-Urban Interface: Lessons for Higher Education. *Choices* 29 (1): 1–6.

Lichter, D. T., and J. P. Ziliak. 2017. The Rural-Urban Interface: New Patterns of Spatial Interdependence and Inequality in America. *Annals, AAPSS* 672: 6–25.

Metropolitan (Met) Council. 2020. *Community Profile for Dakota County*. Metropolitan Council, Minnesota. Accessed May 28, 2020. https://stats.metc.state.mn.us/profile/detail.aspx?c=037#POPANDHH

Mather, M., and A. Lee. 2020. Children Are at the Forefront of U.S. Racial and Ethnic Change. Population Reference Bureau. February 10, 2020. Accessed February 11, 2020. https://www.prb.org/children-are-at-the-forefront-of-u-s-racial-and-ethnic-change/

Meersman, T. 2014. Latinos Are Learning to Farm Minnesota Style. *Star Tribune.* June 9, 2014. Accessed July 7, 2020. https://www.startribune.com/latinos-learning-to-farm-minn-style/262197411/1/

Minnesota Compass. 2020. *All Minnesotans by race and ethnicity.* Minnesota Compass. Accessed May 28, 2020. https://www.mncompass.org/topics/demographics/race-ethnicity?hispanic

Minnesota Compass A. 2020. *Minnesota's Cultural Communities.* Minnesota Compass. Mncompass.org/cultural communities. Accessed May 28, 2020.

Minnesota Grown. 2020. Minnesota Grown Directory. Minnesota Department of Agriculture. Accessed July 1, 2020. https://minnesotagrown.com/search-directory/csa-community-supported-ag-farm/

Metropolitan Council. 2020a. History of the Metropolitan Council. Metropolitan Council. Accessed June 26, 2020. https://metrocouncil.org/About-Us/What-We-Do/History-of-the- Metropolitan-Council.aspx

Metropolitan Council. 2020b. Thrive 2040. Metropolitan Council. Accessed June 26, 2020. https://metrocouncil.org/planning/projects/thrive-2040.aspx.

Pfeffer, M. J., and M. B. Lapping. 1994. Farmland Preservation, Development Rights and the Theory of the Growth Machine: The Views of Planners. *Journal of Rural Studies* 10 (3): 233–248.

Plane, D. A., and P. A. Rogerson. 1994. *The Geographical Analysis of Population: With Applications to Planning and Business.* New York: Wiley.

Pruetz, R., and N. Standridge. 2008. What Makes Transfer of Development Rights Work? Success Factors from Research and Practice. *Journal of the American Planning Association* 75 (1): 78–87.

Robbins, P. 2012. *Political Ecology,* 2 nd ed. Oxford: Wiley-Blackwell.

Taylor, L. 2011. No Boundaries: Exurbia and the Study of Contemporary Urban Dispersion. *GeoJournal* 76(4): 323–339.

The ARDA (The Association of Religion Data Archives). 2020. U.S. Congregational Membership: County Reports. The Association of Religion Data Archives. Accessed June 15, 2020. http://www.thearda.com

Tsu, C. M. 2017. If You Want to Plow Your Field, Don't Kill Your Buffalo to Eat: Hmong Farm Cooperatives and Refugee Resettlement in 1980s Minnesota. *Journal of American Ethnic History* 36 (3):38–72.

U.S. Bureau of Census. 1995. *Population of Counties from Decennial Census: 1900-1990.* Washington, D.C.: US Bureau of Census.

U.S. Bureau of Census. 2000. *Decennial Census 2000.* Washington, D.C.: U.S. Bureau of Census.

U.S. Bureau of Census. 2010. *Decennial Census 2010.* Washington, D.C.: U.S. Bureau of Census.

U.S. Bureau of Census. 2017. *American Community Survey (ACS). 2013 – 2017 American Community Survey estimates.* U.S. Census Bureau's American Community Survey Office. https://factfinder.census.gov/

USDA (United States Department of Agriculture). 2002. *Census of Agriculture.* Washington, D.C.: USDA.

USDA (United States Department of Agriculture). 2007. *Census of Agriculture.* Washington, D.C.: USDA.

USDA (United States Department of Agriculture). 2012. *Census of Agriculture.* Washington, D.C.: USDA.

USDA (United States Department of Agriculture). 2017. *Census of Agriculture.* Washington, D.C.: USDA.

Vang, C. Y. 2008. *Hmong in Minnesota*. Minnesota: Minnesota Historical Society.
Woods, M. 2007. Engaging the Global Countryside: Globalization, Hybridity and the Reconstitution of Rural Place. *Progress in Human Geography* 31 (4): 485–507.
Yin, R. 2009. *Case Study Research: Design and Methods*, 4 th ed. Los Angeles, CA: Sage.

7 Labour immigration and Demographic Transformation

Lithuanian and Polish Nationals in Rural Ireland

Mary Cawley

Introduction

This chapter focuses on increased ethnic diversity as one aspect of demographic transformation in rural areas, using the migration of Lithuanian and Polish nationals into small towns and rural areas in Ireland as an example. The theme answers to Woods's (2012, 2) call for greater understanding of 'new international circuits of migrant labour'. Quantitative methods are used in order to measure change between 2006 and 2016, including choropleth mapping, the application of an index of population concentration and deconcentration, used by Barcus and Simmons (2013) in their research on ethnic diversification in the Great Plains of the United States of America (USA), and an index of dissimilarity (ID) used by Lichter and Johnson (2006) to study segregation in the USA. Further quantitative studies of the distribution of immigrant labour in European countries are recommended by Rye and Slettebak (2020) to complement qualitative case studies. Ireland provides an appropriate context for such a study because of the influx of substantial numbers of labour migrants from the former socialist economies that acceded to membership of the European Union (EU) on the 1st of May 2004 (the Czech Republic, Estonia, Hungary, Latvia, Lithuania, Poland, Slovakia and Slovenia). Polish and Lithuanian nationals are the two largest numerical groups for whom detailed Irish census data are available.

Traditionally, and even yet, immigrants tend to concentrate in gateway cities because of having access to a range of employment. Dispersion down the urban hierarchy takes place but direct movement and dispersal to work in small towns and rural areas have been documented since the late 1990s in Australia, Canada, countries of Europe and the USA, for example (Argent, and Tonts 2015; Barcus, and Simmons 2013; Findlay, and McCollum 2013; Fonseca 2008; Hoggart, and Mendoza 1999; Jentsch, and Simard 2009; Kasimis 2008; Preibisch 2007). The changing demands for labour arising from the restructuring of rural economies in these countries have contributed to this movement. Few opportunities exist for higher-skilled employment in such locations. Immigrants may accept employment below their skills and educational qualifications because even the minimum wage in a developed economy may exceed that available domestically and they may wish to experience living in another country (Castles, de Haas, and Miller 2014). Some immigrants undoubtedly move to highly skilled

DOI: 10.4324/9781003110095-9

and professional positions commensurate with their qualifications. However, in Europe, immigrant labour from other European low wage economies has been identified as often moving to low paid and low skilled employment in rural areas, notably in agriculture, food processing, building and construction, hospitality and tourism, retail, healthcare, domestic work and some forms of non-food manufacturing (Jentsch, and Simard 2009; McAreavey 2017). Labour deficits may exist in these sectors locally, as a result of past outmigration. In addition, an increasingly educated native working-age population is often unwilling to accept the relatively low wage levels and, sometimes, precarious working conditions that are present (Fonseca 2008; Kasimis 2008; Rye, and Scott 2018). Immigrants may therefore serve to fill gaps in local labour markets, as secondary forms of labour, instead of competing directly with a local workforce.

The employment of migrant labour in outdoor horticulture and food production is well documented for Greece, Portugal, Spain and the United Kingdom (UK) (Fonseca 2008; Hoggart, and Mendoza 1999; Kasimis 2008; Rogaly 2008). They also work indoors in, sometimes unhealthy, environments such as mushroom growing units (McAreavey 2012). Food processing can involve demanding unpleasant work in abattoirs, meatpacking and fish processing facilities (Rye, and Scott 2018). Large-scale civic construction projects are often a major source of short-term employment for migrant males (Chan, Clarke, and Dainty 2010). Seasonal labour demands in restaurants and hotels in tourist areas are frequently filled by migrant workers, and they may work unsocial hours that are not acceptable to local workers in retail establishments and cafés (Rye, and Scott 2018). Migrant female workers may find employment in elderly care homes and as cleaners in private homes (Kasimis 2008). In some countries, migrants are employed in certain forms of manufacturing activity, such as textiles in Portugal (Fonseca 2008). Migrant workers contribute to the maintenance of rural economies and to the diversification of rural populations.

The aims of the research discussed here are: (i) to establish the spatial distribution of Lithuanian and Polish migrant workers at a range of geographical scales and (ii) to identify the implications for ethnic diversification in rural Ireland. It is hoped to contribute to better understanding of labour immigration in rural Ireland and to add to the international literature on this theme. The chapter provides background context for Ireland, describes the methodology and presents the results, followed by a conclusion.

Context: Migrant Labour Immigration in Ireland

The current Irish state was characterized by emigration from the 1840s until 1991, particularly, but not exclusively from rural areas, because of the inability of the economy to absorb the natural increase in population (Sexton et al. 1991). In the early 1990s, Ireland became a country of immigration, and the citizens of continental European countries formed an increasing component of the total, in both urban and rural areas (CSO 2012a). Two main factors have contributed to this inflow. First, an unmet domestic demand for labour arose during increased economic growth from the early 1990s and, especially, between 1998 and a deep recession, beginning in 2008 (linked to the international financial crisis but precipitated by domestic factors) (Krings et al. 2009). Second, the enlargement of the EU, on the 1st of May 2004, to admit the eight former Soviet Bloc countries to full membership, provided a new source of labour and Ireland (with Sweden and the UK) granted immediate access to its labour market (Quinn 2010). Many of the new immigrants were unable to find satisfactory employment in their own countries (Favell 2008). Demand for low-skill labour continued after 2008 and the return of Polish and Lithuanian migrants to the areas of origin was less than might have been anticipated (Krings et al. 2011).

Just over 2000 Lithuanian and Polish nationals, respectively, were recorded as being residents in the 2002 Irish census, but an unknown amount of under-reporting took place (Table 7.1). The major increase in numbers between 2002 and 2006 is therefore considered to be an overestimate, although large numbers of both groups moved to Ireland after 2004 (MacÉinrí and White 2008). Dublin, the capital city, and Cork, Limerick, Galway and Waterford cities were the main initial destinations, but migration took place also to towns of various sizes and rural areas (Table 7.2). The CSO has identified the sectors in which the migrants found employment at an aggregate level as building and construction, manufacturing, wholesale and retail activities, hotels and restaurants, and business services (CSO 2008, 30 and 34). Major growth occurred in both populations between 2006 and 2011 (mainly until 2009) and minor declines occurred in both between 2011 and 2016, a period of continued recession and recovery (Table 7.1). There was increased evidence of family formation in the latter period (CSO 2017a). The proportion of children of foreign nationality under five fell between 2011 and 2016, in large part because of an increasing number of children born to a Polish or Lithuanian parent or parents being registered as having dual Irish–Polish and Irish–Lithuanian nationality (CSO 2017a).

The largest proportions of Lithuanian and Polish immigrants resided in major cities and towns in all three census years (Table 7.2). Nevertheless, as the aggregate census data show, movement took place to smaller towns and rural areas from an early stage to fill vacancies in particular economic sectors. These

Table 7.1 Ireland: population usually resident and present on census night, by nationality

Population	2002	2006	2011	2016	% Change		
					2002– 2006	2006– 2011	2011– 2016
State total	3,744,059	4,172,013	4,525,281	4,689,921	11.43	8.47	3.64
N Lithuanian	2,104	2,4628	36,683	36,552	1,070.53	48.95	−0.36
% Lithuanian	0.05	0.60	0.81	0.78			
N Polish	2,091	63,276	122,585	122,515	2,926.11	93.73	−0.06
% Polish	0.05	1.52	2.71	2.61			

Sources: CSO (2003), Table B0439; CSO (2007a), Table C0437; CSO (2017a), Table E2070.
Note: The population resident and present on census night is slightly less than the total population. This definition is used by the Central Statistics Office in the small area population statistics for nationality groups.

Table 7.2 Irish, Lithuanian and Polish nationals: % distribution between different town size groupings and rural areas

	2006 (%)			2011 (%)			2016 (%)		
	Irish	Lithuanian	Polish	Irish	Lithuanian	Polish	Irish	Lithuanian	Polish
Dublin City and suburbs	23.52	26.70	24.89	22.89	24.89	24.95	23.25	22.72	22.75
Other cities and suburbs	9.22	8.00	15.8	8.90	7.84	14.03	8.84	7.25	13.13
Towns 10,000 and over	14.01	24.7	23.2	15.18	29.14	27.64	15.70	31.07	28.89
Towns 5,000–9,999	6.29	14.8	11.1	6.21	13.43	13.11	5.92	13.76	13.20
Towns 3,000–4,999	2.49	5.90	4.8						
Towns 1,500–2,999	2.94	5.60	4.4						
Rural areas <1,500 population	41.53	14.30	12.3						
Towns 2,000–4,999				4.70	10.38	8.59	4.70	10.13	8.67
Towns 1,500–1,999				1.43	2.10	1.83	1.55	2.05	2.15
Towns 1,000–1,499				2.07	2.59	1.98	2.06	2.36	2.06
Towns 500–999				2.79	2.15	2.16	2.87	2.14	2.27
Towns <500				2.67	1.36	1.27	2.60	1.45	1.24
Remainder of country				33.20	6.13	4.44	32.50	6.80	5.64
Total number	3,613,498	24,628	62,674	392,7143	36,683	122,585	4,082,513	36,552	122,515

included beef processing, 80% of which is exported, which had grown markedly during the late 1990s and is located in small and medium-sized towns throughout the country (Crowley, Walsh, and Meredith 2008; Maher and Cawley 2016). Lamb, pig and poultry processing also increased, with the latter two being concentrated, respectively, in the north midlands' counties of Cavan and Monaghan (Figure 7.1) (Crowley et al. 2008, 37). Ireland became a major producer of

Figure 7.1 Ireland: Counties.

Author, based on Irish Central Statistics Office shapefiles.

mushrooms during the 1980s with an initial concentration in small towns in county Monaghan, although more widely distributed later (Crowley et al. 2008, 211). Building and construction was a key contributor to employment growth during 1998–2008, involving private housing and large-scale road construction projects, and Lithuanian and Polish men were recruited in substantial numbers (Krings et al. 2011). Speculative housing construction took place in many small towns and villages, much of which remained vacant following the recession. Following the onset of recession in 2008, unemployment reached 60% among the EU migrant workers, and some migrants returned home (Krings et al. 2009). Population growth created increased demand for food and the expansion of horticultural production in a traditional vegetable growing area north of Dublin, in the environs of other large cities and in parts of the northwest and the southeast (Crowley et al. 2008, 214–215). The rapidly expanding tourism sector in scenic rural areas, the expansion of restaurants and cafés and retail establishments in small and large settlements, during the years of economic growth, also provided employment, particularly for female immigrants (Wickham et al. 2008).

Case studies of the Lithuanian and Polish immigrant experience in Ireland reveal that, whilst satisfaction was expressed by some with increased incomes, improved personal circumstances and the experience of living in a different country, problems existed which include inadequate recognition of qualifications and poor working conditions (Coakley, and MacÉinrí 2009; Gilmartin, O'Connell, and Migge 2008). Inadequate English language skills resulted in difficulties in accessing employment and poor understanding of labour rights and led to discrimination in the workplace (Coakley, and MacÉinrí 2009, 113). Precarious working conditions were identified in the mushroom sector in the north midlands (Arqueros-Fernández 2009). Dependence on project work with short-term contracts and pressure on wages were common in the construction sector (Krings et al. 2011).

Data Sources and Methods of Analysis

Following Barcus, and Simmons (2013) the distribution of the Polish and Lithuanian immigrants and change in the distribution over time were mapped. The Hoover Index (HI) was used to measure the extent of concentration and deconcentration of the populations (Rogerson, and Plane 2013) and the ID was applied to measure the degree of segregation of the immigrants from the majority Irish population (Lichter, and Johnson 2006). Cluster analysis was applied, in order to identify if associations existed between the distribution of the immigrant groups and particular sectors of employment by area. Pearson correlation values were calculated to measure the strength of the relationships.

The study is based on published and unpublished data for various geographical areas for the census years 2006, 2011 and 2016. Population counts are available for Lithuanian and Polish nationals but not for other recent immigrant groups because of small numbers and issues relating to confidentiality. Employment data for the total populations of the Electoral Districts (EDs) are used in the cluster analysis; disaggregation by nationality is not available. Numbers of establishments are based on membership lists. Nationality, rather than place of birth, was selected to define the populations, in order to include children with foreign nationality who were born in Ireland. Study of only two nationality groups gives a somewhat limited view of the changing ethnic composition of the population in the Irish countryside since 2004, but Poles and Lithuanians are the two largest recent EU immigrant groups, comprising 59.6% and 17.8%, respectively, of the total 205,398 post-2004 accession state nationals in 2016 (CSO 2017a, Table E7002). They illustrate the establishment and spread of new labour immigrants in a country over a relatively short period of time which includes a period of major economic growth, the recession of 2008, and years of recovery between approximately 2012 and 2016.

The study analysed national patterns initially for 31 census county divisions and the 3,409 EDs, as contexts against which urban and rural differences could be identified. Attention then focused on 'rural' areas, as defined for the purposes of study. There are 26 counties in the Irish state and 31 census divisions: Dublin County consists of four divisions, and both Cork and Galway contain City and County divisions. Counties are further divided into Aggregate Town Areas (ATAs), with a population of 1,500 and over and Aggregate Rural Areas (ARAs), with a population of less than 1,500 and including open countryside. The HI and the ID were applied at county, ATA and ARA scales.

The 3,409 EDs include urban and rural areas. The EDs vary in size from less than one square kilometre to more than 100 square kilometres and the populations vary from several thousands in urban areas to less than 100 in some peripheral rural locations. Choropleth mapping was used to illustrate the density per 10 square kilometres and the distribution of Lithuanian and Polish nationals by the ED, for each of the three census years. This mapping permits urban and rural distributions to be identified visually. For reasons of space only the density maps for 2016 are included here, but the main patterns and trends are discussed. The HI and the ID were also calculated for the EDs for each census year. Because parts of the environs of the urban EDs overlap adjoining rural EDs, further disaggregation of the EDs was undertaken to attain a better understanding of trends in the more rural areas in 2016.

The largest concentrations of Lithuanian and Polish nationals in Ireland are in Dublin and the other four large cities and in 47 towns (including their environs) with populations of 10,000 and over (Table 7.2). In order to identify less urbanized EDs, these settlements were excluded from the database, following detailed examination of the distribution of the town environs in an unpublished census file, and the remaining districts with smaller populations were considered 'rural'. The rubric adopted for removing an ED from the file was that in excess of 50% of its population was in the environs of a town of 10,000 population or over. When the environs were included, some 599 EDs were classified as urban and removed from the file, leaving the 2,810 EDs classified as 'rural'. In most cases, the percentage in a town's environs in the adjoining EDs exceeded 70%. The HI and the ID were calculated for these EDs for 2016. In order to apply cluster analysis to identify the association between the distribution of the immigrant groups and sectors of employment, the EDs which contained neither Lithuanian nor Polish nationals in 2016 were removed from the file, giving the 1,903 EDs for clustering.

The HI measures the extent of concentration and deconcentration of a population in a region that is disaggregated into a set of subregions (counties, ATAs, ARAs and the EDs, in this instance). The index can range from 0 to 100 with the larger values representing a higher degree of concentration. The value of the index can be interpreted as the proportion of the total population that would need to be redistributed across subregions to achieve equal population densities in all subregions (Rogerson, and Plane 2013, 99). A decrease in the value of the index over time would indicate that the population in question is becoming more dispersed and an increase would indicate greater concentration. The index is calculated as follows:

$$H_t = \frac{1}{2} \sum_{i=1}^{n} |p_{it} - a_i|$$

Where p_{it} and a_i denote subregion i's share of the total population at time t and its area, respectively, and where there are n subregions.

A second measure used by Lichter and Johnson (2006) to assess ethnic segregation in the USA, the ID, measures the relative distribution of one population in relation to another across geographical areas. It is used here to measure dissimilarity between the distribution of the Lithuanian and Polish immigrant groups versus the Irish population. Like the HI, the value of the ID can range from 0 to 100. The value indicates the percentage of a minority population that would need to be redistributed to be similar in distribution to the majority comparator population. The index is expressed as follows:

$$D_t = \frac{1}{2} \sum_{i=1}^{n} |m_{it} - c_{it}|$$

Where m_{it} and c_{it} are the respective percentages of a minority migrant group (Lithuanian or Polish nationals) and the comparator population (Irish nationals) residing in an ED i at time t. If the minority population percentage (m_{it}) and the comparator percentage are equal in all the EDs then the index is equal to 0 meaning that they are distributed in the same percentages over all the EDs and residential segregation is low (Lichter, and Johnson 2006, 116). If the index equals 100 this means that segregation is high and 100% of the minority population would have to move to the other EDs to be similarly distributed to the majority population. The ID values serve to complement the information available from the HI.

Cluster analysis, using a routine from the Statistical Package for the Social Sciences (IBM Corp., 2019), was applied also to seek to identify if a relationship existed between the distribution of the Lithuanian and Polish nationals and specific employment sectors. The analysis was applied to the reduced file of the 1,903 EDs described above. Eleven variables of four types were used for clustering: economic, demographic, a degree of urbanization variable and indicators of specialist production and processing activities. The percentage of the total employed ED population in five industry types, in which Lithuanian and Polish nationals are identified as working by the CSO (2008), was included: agriculture, forestry and fishing, building and construction, manufacturing, commerce and trade (including wholesale and retail activities) and 'other' (the hospitality, hotel and restaurant sector, social and personal services). Two variables were related to the percentage of Polish and Lithuanian nationals in each ED. Four binary variables were included: the presence or absence of a medium-size town of 1,500–9,999 population and, to cover some of the principal activities engaged in, the presence or absence of a meat processing plant, a mushroom growing plant and year-round horticultural operations. The values were standardized before clustering, using z values, and hierarchical cluster analysis was applied using Ward's method. Pearson correlation was applied to measure the strength of the relationship between nationality and the other variables.

Results

The Distribution of Lithuanian and Polish Nationals, 2006–2016

Dublin City, the capital, and its suburbs and Cork, Limerick and Galway cities, and Waterford to a lesser extent, were major destinations for the Lithuanian and Polish immigrant workers (Table 7.2). The relative percentages of both nationality groups (approximately 25%) exceeded that of the Irish national population in the capital area until 2016. Lithuanians moved to smaller places earlier and in slightly higher percentages than did Polish nationals, who were attracted to other cities and their suburbs to a greater extent. There is evidence of movement towards towns with a population of 10,000 and over which contain county capitals and other large settlements (in Irish terms) among both immigrant groups and Irish nationals. By contrast, the relative percentage of the immigrants (between 11% and 15%) in towns with 5,000–9,999 population was double that of the Irish nationals in all three years, suggesting the adoption of employment that may have been less attractive to Irish workers. In 2006, Lithuanian and Polish nationals were less represented than Irish nationals, at an aggregate level, in places with less than 1,500 population. Data for small towns for 2011 and 2016, however, illustrate movement of the immigrants to towns of 2,000–4,999 population (Table 7.2). Low percentages of both immigrant groups, but more Lithuanians than Poles, were present in smaller places and areas of open countryside. Possible factors that contributed to this movement, after 2011, include family formation and the birth of children, as noted earlier, the cheaper cost of housing, because of excess supply, and an improving economy which provided employment.

The distribution of Lithuanian and Polish nationals was mapped as a percentage of the ED population and as a density per 10 square kilometres for the three census years. Because of limitations of space, only the density maps for the two groups are presented for 2016, and they illustrate the maximum spread of population to date. Because of their larger numbers, Polish nationals have been present in a larger number of the EDs than Lithuanians since 2006. The maps not presented here, as might be expected from Table 7.1, illustrate an association of both groups with cities, towns and adjoining rural areas. The densities decreased with distance from towns with a population of 10,000 and over. Increased move- ment into more rural areas was apparent but, in 2016, no Lithuanians were res- ident in 1,896 of the EDs (56.5% of the total) and no Polish were resident in 1,004 of the EDs (29.5% of the total) (Figures 7.2 and 7.3). These coincide with

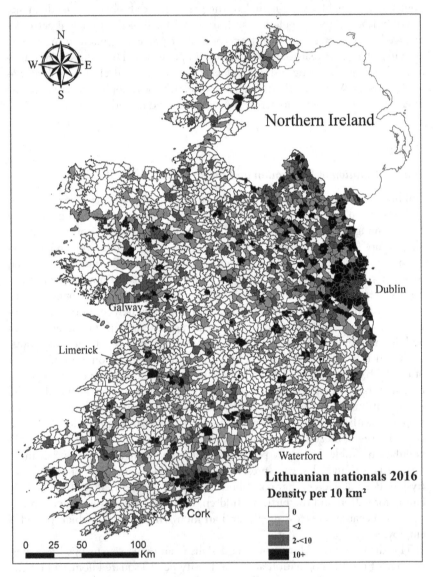

Figure 7.2 Ireland: distribution of Lithuanian nationals, 2016.

Author, based on Irish Central Statistics Office data and shapefiles.

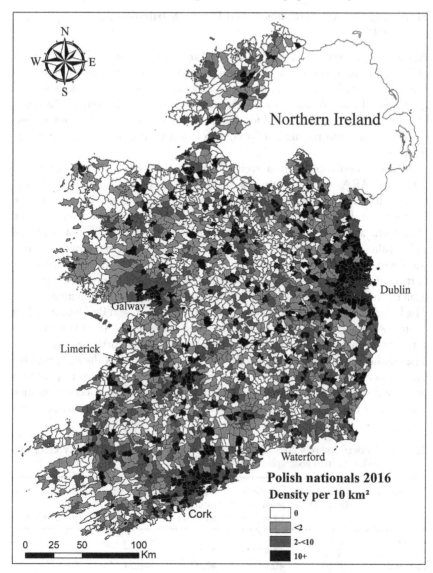

Figure 7.3 Ireland: distribution of Polish nationals, 2016.

Author, based on Irish Central Statistics Office data and shapefiles.

upland environments in Atlantic coastal areas and County Wicklow, south of Dublin, and in agricultural areas outside urban influence.

Apart from differences in the greater spatial extent and density of Polish than of Lithuanian nationals throughout much of Ireland, the latter show a stronger association with the northeast of the state, arising from their involvement in horticulture in counties Dublin and Meath and in mushroom production and poultry processing in Monaghan and Cavan, respectively (Figures 7.1 and 7.2). Over time, Polish nationals have also moved into employment in these areas.

Measuring Changing Lithuanian and Polish Distributions, 2006–2016

At the level of town size groupings and rural areas, Lithuanian and Polish nationals moved from the larger gateway cities to smaller places over time (Table 7.2). The HI permits this apparent deconcentration to be examined at a more detailed geographical scale. Analysis was conducted for counties, ATAs and ARAs within counties, for the 3,409 EDs and for the 2,810 'rural' EDs. The HI values are presented for the Irish population for comparison with the two immigrant groups in Table 7.3.

All three populations were unevenly distributed between counties and between ATAs and ARAs within counties in all three years (Table 7.3). The Irish population differed from the two immigrant groups in being less concentrated at a county level and, notably, in ARAs within counties, as one would expect. Deconcentration to ARAs increased substantially between 2006 and 2011 (the index value fell from 23.91 to 10.47), indicative of residential movement into the countryside (Gkartzios and Scott 2010). Nevertheless, the substantial proportion of the native population that resides in cities and large towns is apparent from the higher index values for ATAs in all years than for the immigrant groups. The broad similarities in the index values for Lithuanian and Polish nationals at county, ATA and ARA scales in the three years (between 44.48 and 49.86) indicate considerable concentration in particular urban and particular rural areas. The smaller-scale 3,409 EDs include cities, towns and rural districts, and the index values underline the concentration of all three populations in particular urban and rural areas at this more detailed geographical scale. Deconcentration was greatest among Polish nationals between 2006 and 2011 (a change in the

Table 7.3 Hoover Index values of population concentration of nationality groups for different Irish geographical areas, 2006, 2011 and 2016

Nationality	Counties (n = 31)	Aggregate town areas (n = 31)	Aggregate rural areas (n = 31)	Electoral districts (n = 3,409)	'Rural' electoral districts (n = 2,810)
2006					
Irish	34.60	52.88	23.91	53.77	
Lithuanian	49.70	49.58	49.90	49.74	
Polish	49.33	49.23	49.76	53.22	
2011					
Irish	34.23	51.02	10.47	52.79	
Lithuanian	49.59	48.48	49.65	49.66	
Polish	48.81	49.42	49.87	49.25	
2016					
Irish	34.91	51.50	10.52	53.58	34.18
Lithuanian	48.77	48.45	49.61	49.65	49.73
Polish	49.61	48.45	49.86	49.15	49.23

Sources: Counties: estimates calculated from Small Area Population Statistics, 2006 (CSO 2007b); special tabulations provided by the Irish Central Statistics Office, 2011, 2016. Electoral Districts: CSO (2007b, 2012b, 2017b).

index from 53.22 to 49.25). There is also evidence of minor deconcentration among Lithuanian and Polish nationals over the two census periods.

At the level of the 2,810 'rural' EDs, where cities and large towns are excluded, the HI values reveal greater dispersal of Irish nationals among such areas but concentration of both Lithuanian and Polish nationals with that of the latter being slightly less marked than that of the former (Table 7.3).

As a measure of dissimilarity, the ID measures the distribution of one pop-ulation in relation to another population and indicates the percentage of the minority group that would have to be redistributed in order to have a similar distribution to the majority group (Irish nationals). The index values provide further insight into the HI values which measure the distribution of each group by density. At the scale of ATAs and ARAs, both immigrant groups were more concentrated in particular towns and rural areas than were Irish nationals in all years (Table 7.4). In all except two cases (Polish nationals in ATAs in 2011 and 2016), over 40% of the immigrant group would have to be redistributed in both town and rural areas to be similarly distributed to Irish nationals. The extent of redistribution required was less in the case of Polish than of Lithuanian nationals in all years, pointing to slightly less dissimilarity from Irish nationals (Table 7.4). On the scale of 3,409 EDs, the index values were lower but again more than 40% of the immigrant groups would have to be redistributed to be similarly distributed to Irish nationals. The index values declined marginally, in most instances, over time and were lower in the case of Polish nationals than of Lithuanians. When cities and towns with populations in excess of 10,000 were excluded (the 2,810 EDs in 2016), the indices were slightly higher than in the case of the 3,409 EDs, indicating greater concentration in specific rural EDs than obtained among Irish nationals.

The ID values reveal that considerable disparities exist between the distribu-tion of the two immigrant groups and Irish nationals within both urban and rural areas. Lithuanian and Polish nationals, and particularly the latter, are present in many EDs but they remain minorities in relation to Irish nationals.

Table 7.4 Index of dissimilarity values for Irish nationals and Polish and Lithuanian nationals for different geographical areas, 2006, 2011 and 2016

Nationality	Counties (n = 31)	Aggregate Town Areas (n = 31)	Aggregate Rural Areas (n = 31)	Electoral Districts (n = 3,409)	'Rural' Electoral Districts (n = 2,810)
2006					
Irish vs. Lithuanian	44.70	44.81	49.20	44.13	
Irish vs. Polish	44.23	44.13	49.09	43.66	
2011					
Irish vs. Lithuanian	42.98	40.94	46.31	42.99	
Irish vs. Polish	42.04	39.55	46.08	42.04	
2016					
Irish vs. Lithuanian	43.13	41.17	46.42	43.13	45.08
Irish vs. Polish	42.22	39.86	46.16	42.22	44.45

Relationships between the Distribution of Lithuanian and Polish Nationals and Employment

Cluster analysis was performed to gain insight into the association, if any, between the distribution of Lithuanian and Polish nationals in 1,903 'rural' EDs, in 2016, and particular sectors of employment. The strength of the relationships was measured using Pearson correlation; coefficient values that are significant at the 0.001 level are referred to. Four clusters were selected to partition the data, which contain, respectively, 231, 1,603, 29 and 40 EDs (12.1%, 84.2%, 1.5% and 2.1% of the total) (Table 7.5). Cluster 1 is a 'town' cluster with a presence of meat processing facilities; Cluster 2 consists of the remoter rural EDs; Cluster 3 has towns in almost 60% of its EDs and a concentration of non-mushroom horticultural production; and Cluster 4 contains towns in 50% of its EDs and a concentration of mushroom production units. The clusters are mapped in Figure 7.4.

The reduced importance of employment in construction after 2008 may be noted at the outset; a significant positive relationship (a coefficient of +0.505) was recorded only in the case of Lithuanians in Cluster 3 (Table 7.5). The Cluster 1 EDs are distributed in small numbers across all counties, except county Leitrim (Figure 7.1). The presence of towns in most EDs provides a range of employment in manufacturing, wholesale and retail activities and 'other' (hospitality,

Table 7.5 Results of cluster analysis of 'Rural' electoral districts: average variable values

Variables	Cluster 1 Towns/ Meat Processing (n = 231)	Cluster 2 Remote Rural (n = 1603)	Cluster 3 Horticulture/Towns (n = 29)	Cluster 4 Mushroom/ Towns (n = 40)	Average (1,903 EDs)
Agriculture, forestry, fishing	8.615	13.282	10.412	10.989	12.628
Building and construction	5.961	6.781	5.767	6.320	6.642
Manufacturing industry	14.728	13.077	14.189	13.283	13.307
Commerce and trade	22.283	19.247	21.904	21.596	19.713
'Other'	17.691	14.153	16.821	16.229	14.676
% Lithuanian nationals	1.253	0.275	0.913	1.474	0.429
% Polish nationals	3.531	0.873	3.088	2.617	1.268
Town of 1,500–9,999 population (N)	208	3	17	20	19
Meat processing plant (N)	41	1	4	2	
Mushroom production unit (N)	1	0	5	39	
Other horticultural production (N)	0	0	29	0	

Sources: Census 2016, Small Area Population Statistics (CSO 2017b) and registered producer lists.

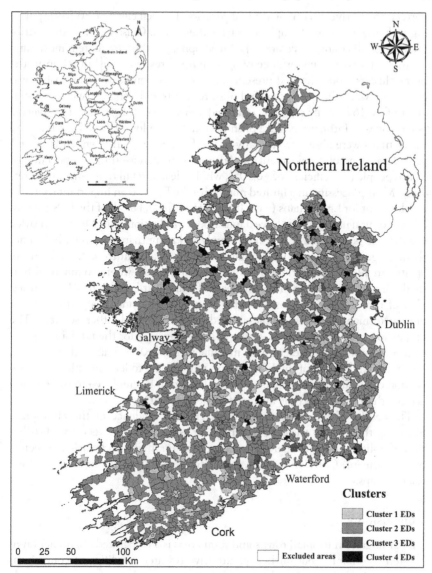

Figure 7.4 Ireland: distribution of clusters of EDs.

Author, based on Irish Central Statistics Office data and shapefiles.

social and personal) services. The cluster has the highest representation of Polish nationals and the second highest representation of Lithuanian nationals among the 1,903 EDs. Significant relationships were present between the percentages of Polish and Lithuanian residents and both manufacturing (+0.460 and +0.245, respectively) and 'other' services (+0.215 and +0.177, respectively). The presence of 41 meat processing plants is linked to manufacturing employment, being classified as such in the Irish census. The percentage of Lithuanian residents also correlated positively with meat plant location (a coefficient of +0.246), pointing to being a source of employment for them. The Cluster 2 EDs are distributed across all counties (Figure 7.4). Geographically they are more remote rural areas, with a greater importance of agriculture, forestry and fishing than in the other clusters, and a limited presence of towns. A significant relationship was present between the percentage of Polish residents and 'other' services (a coefficient of +0.165). Only one ED contained a meat processing plant. The average percentages of Lithuanian and Polish nationals were lowest for all clusters, and Lithuanians were absent from many of the EDs. Seventeen towns in Cluster 3 provide a range of employment opportunities in manufacturing, especially for Polish people (a coefficient of +0.505), wholesale and retail trade and 'other' services. Meat processing in a limited number of the EDs was an important source of employment for Lithuanians (a coefficient of +0.743), but all of the EDs contain a horticultural unit and five have a mushroom production unit. More than twice the average percentage of Polish nationals and more than one and a half times the average percentage of Lithuanian nationals resided in these EDs. They are located in 14 counties, primarily in the east and south, where commercial horticulture is most practised. The Cluster 4 EDs are distributed across 19 counties. Half of the EDs contain towns and a range of employment opportunities are present in manufacturing, wholesale and retail trade, and 'other' services. The proportion of Polish nationals is over twice the average for the total and significant relationships are present with manufacturing (+0.546) and 'other' services (+0.335). All but one of the EDs contain mushroom production units. The proportion of Lithuanian nationals (who are known to be employed in these units) is more than three times the average for the 1,903 EDs.

The results of the cluster analysis extend the evidence of the choropleth mapping, the HI and the ID by identifying associations between the distribution of Polish and Lithuanian nationals in the 'rural' EDs and particular sectors of employment. Correlation analysis illustrates the strengths of some of these relationships.

Conclusion

Labour migration to small towns and rural areas is a recognized phenomenon in Western economies, but further quantitative research for different countries is recommended to expand the knowledge base and provide contexts for case study research (Rye, and Slettebak 2020). This chapter documented the experience of Polish and Lithuanian immigrants who were attracted to Ireland during a period of rapid economic growth, notably following enlargement of the EU in May 2004, when they had immediate access to the Irish labour market. The research followed broadly a methodology used by Barcus and Simmons (2013) in the USA and used published and unpublished census data at a range of geographical scales. The migrants moved to large cities and towns and also to smaller towns and rural areas to meet an unmet labour demand, among which construction activities were of particular importance for males before the recession of 2008. Cluster analysis and correlation analysis, based on 2016 data, illustrate that construction had declined in importance for the most part. There was a close association between the distribution of the larger numbers of Polish nationals with manufacturing and a range of social and personal services on a widespread scale throughout the state. The distribution of the less numerous Lithuanian nationals aligned more closely with meat processing facilities, mushroom and other horticultural production, particularly in north midland and northeastern counties. The HI values illustrated concentration of both groups in particular urban and rural areas and dispersion to smaller places as the economy recovered from recession from 2012 on.

The numbers involved in an ED varied according to the source of employment and were small in many cases. However, important contributions have been made to filling vacancies in sectors that produce and process local resources, provide services for tourists and local people, including the elderly in care homes, advancing rural economic development and strengthening local demographic structures. The ID values of dissimilarity, which compared the relative distribution of the two immigrant groups with Irish nationals, illustrate the growing ethnic diversity arising from the immigration of Polish and Lithuanian nationals (and other immigrant groups, not studied here) in rural Ireland. They remain minorities in most EDs and limited numerical transformation is taking place, except in towns where meat processing, mushroom production or particular social and personal services are located. Nevertheless, new transnational links are being established between many towns, villages and rural areas in Ireland and urban and rural places in Poland and Lithuania (Woods 2007, 2012). The full implications of the ethnic restructuring that is taking place in small towns and rural Ireland, arising from Lithuanian and Polish immigration, require further detailed analysis of age and family structures at a localized scale and exploration of the dynamics of the relationships between the immigrant groups and Irish people.

Acknowledgements

My thanks are due to the Research Support Fund at NUI Galway for funding the GIS mapping and to Mary Callaghan who prepared Figures 7.2–7.4, from the prepared data. Dr Siubhán Comer is thanked for Figure 7.1. Professor Serge Schmitz and Lauriano Pepe of the Laplec laboratory at the University of Liège are acknowledged for their assistance in preparing the maps for publication. The statistics are copyright of the Government of Ireland and they and the accompanying boundaries, provided for general information by the Central Statistics Office, are accessible free of charge and licensed under Creative Commons Attribution (version 4.0 cc-by).

References

Acqueros-Fernández, F. 2009. Contrasts and Contradictions in Union Organising: The Irish Mushroom Industry. In *The Future of Union Organising*, ed. G. Gall, 205–222. London: Palgrave Macmillan.

Argent, N., and M. Tonts. 2015. A Multicultural and Multifunctional Countryside? International Labour Migration and Australia's Productivist Heartlands. *Population, Space and Place* 21: 140–156.

Barcus, H. R., and L. Simmons. 2013. Ethnic Restructuring in Rural America: Migration and the Changing Faces of Rural Communities in the Great Plains. *The Professional Geographer* 65 (1): 130–152.

Castles, S., H. de Haas, and M. J. Miller. 2014. *The Age of Migration*. Basingstoke: Palgrave Macmillan.

CSO (Central Statistics Office). 2003. *Census 2002, Volume 4, Usual Residence, Migration, Birthplaces and Nationalities*. Dublin: Stationery Office.

CSO (Central Statistics Office). 2007a. *Census 2006, Volume 4, Usual Residence, Migration, Birthplaces and Nationalities*. Dublin: Stationery Office.

CSO (Central Statistics Office). 2007b. *Census 2006, Small Area Population Statistics*. https://www.cso.ie/en/census/census2006smallareapopulationstatisticssaps/

CSO (Central Statistics Office). 2008. *Census 2006, Non-Irish Nationals Living in Ireland*. Dublin: Stationery Office.

CSO (Central Statistics Office). 2012a. *Census 2011, Profile 6, Migration and Diversity in Ireland – A Profile of Diversity in Ireland*. Dublin: Stationery Office. https://www.cso.ie/en/media/csoie/census/documents/census2011profile6/Profile_6_Migration_and_Diversity_entire_doc.pdf

CSO (Central Statistics Office). 2012b. *Census 2011, Small Area Population Statistics*. https://www.cso.ie/en/census/census2011smallareapopulationstatisticssaps/

CSO (Central Statistics Office). 2017a. *Census 2016, Profile 7, Migration and Diversity in Ireland*. https://www.cso.ie/en/releasesandpublications/ep/p-n_and_Diversity_entire_doc.pdfcp7md/p7md/p7anii/

CSO (Central Statistics Office). 2017b. *Census 2016, Small Area Population Statistics, Electoral Divisions*. Accessible at: https://www.cso.ie/en/census/census2016reports/census2016smallareapopulationstatistics/

Chan, P., L. Clarke, and A. Dainty. 2010. The Dynamics of Migrant Employment in Construction: Can Supply of Skilled Labour Ever Meet Demand?. In *Who Needs Migrant Workers?* ed. M. Ruhs, and B. Anderson, 225–255. Oxford: Oxford University Press.

Coakley, L., and P. MacÉinrí. 2009. Migration to Rural Ireland: A North Cork Case Study. In *International Migration and Rural Areas – Cross-National Comparative Perspectives*, ed. B. Jentsch, and M. Simard, 99–126. Farnham: Ashgate.

Crowley, C., J. Walsh, and D. Meredith. 2008. *Irish Farming at the Millennium: A Census Atlas*. Maynooth: National Institute for Regional and Spatial Analysis.

Favell, A. 2008. The New Face of East-West Migration in Europe. *Journal of Ethnic and Migration Studies* 34 (5): 701–716.

Findlay, A., and D. McCollum. 2013. Recruitment and Employment Regimes: Migrant Labour Channels in the UK's Rural Agribusiness Sector, from Accession to Recession. *Journal of Rural Studies* 30: 10–19.

Fonseca, M. L. 2008. New Waves of Immigration to Small Towns and Rural Areas in Portugal. *Population, Space and Place* 14 (6): 525–535.

Gilmartin, M., J. A. O'Connell, and B. Migge. 2008. Lithuanians in Ireland. *Oikos: Lithuanian Migration and Diaspora Studies* 5 (1): 49–62.

Gkartzios, M., and M. Scott. 2010. Residential Mobilities and House Building in Rural Ireland: Evidence From Three Case Studies. *Sociologia Ruralis* 50 (1): 64–84.

Hoggart, K., and C. Mendoza. 1999. African Immigrant Workers in Spanish Agriculture. *Sociologia Ruralis* 39 (4): 538–562.

IBM Corp. (2019). *Statistical Package for the Social Sciences, Statistics for Windows, Version 26*. Amonk, New York: IBM Corp.

Jentsch, B., and M. Simard. 2009. *International Migration and Rural Areas: Cross-national Comparative Perspectives*. Farnham Surrey: Ashgate.

Kasimis, C. 2008. Survival and Expansion: Migrants in Greek Rural Regions. *Population, Space and Place* 14 (6): 511–524.

Krings, T., A. Bobek, E. Moriarty, J. Salamońska, and J. Wickham. 2009. Migration and Recession: Polish Migrants in Post-Celtic Tiger Ireland. *Sociological Research Online*, 14 (2): 111–116.

Krings, T., A. Bobek, E. Moriarty, J. Salamońska, and J. Wickham. 2011. From Boom to Bust: Migrant Labour and Employers in the Irish Construction Sector. *Economic and Industrial Democracy* 32 (3): 459–476.

Lichter, D. T., and K. M. Johnson. 2006. Emerging Rural Settlement Patterns and the Geographic Distribution of America's New Immigrants. *Rural Sociology* 71 (1): 109–131.

Maher, G., and M. Cawley. 2016. Short-term Labour Migration: Brazilian Migrants in Ireland. *Population, Space and Place* 22 (2): 23–35.

MacÉinrí, P., and A. White. 2008. Immigration into the Republic of Ireland: A Bibliography of Recent Research. *Irish Geography* 41 (2): 151–179.

McAreavey, R. 2012. Resistance or Resilience? Tracking the Pathway of Recent Arrivals to a 'New' Rural Destination. *Sociologia Ruralis* 52 (4): 488–507.

McAreavey, R. 2017. *New Immigration Destinations: Migrating to Rural and Peripheral Areas*. Abingdon: Routledge.

Preibisch, K. L. 2007. Local Produce, Foreign Labor: Labor Mobility Programs and Global Trade Competitiveness in Canada. *Rural Sociology* 72 (3): 418–449.

Quinn, E. 2010. *Satisfying Labour Demand Through Migration*. Dublin: Economic and Social Research Institute and European Migration Network.

Rogaly, B. 2008. Intensification of Workplace Regimes in British Horticulture: The Role of Migrant Workers. *Population, Space and Place* 14 (6): 497–510.

Rogerson, P. A., and D. A. Plane. 2013. The Hoover Index of Population Concentration and the Demographic Components of Change: An Article in Memory of Andy Isserman. *International Regional Science Review* 36 (1): 97–114.

Rye, J. F., and S. Scott. 2018. International Labour Migration to/in Rural Europe: A Review of the Evidence. *Sociologia Ruralis* 58 (4): 928–952.

Rye, J. F., and M. H. Slettebak, 2020. The New Geography of Labour Migration: EU11 Migrants in Rural Norway. *Journal of Rural Studies* 75: 125–131.

Sexton, G., B. M. Walsh, D. F. Hannan, and D. McMahon. 1991. *The Economic and Social Implications of Emigration, Report No. 90*. Dublin: National Economic and Social Council.

Wickham, J., E. Moriarty, A. Bobek, and J. Salamońska. 2008. *Migrant Workers and the Irish Hospitality Sector*. Dublin: Employment Research Centre and Trinity Immigration Initiative, Trinity College Dublin.

Woods, M.. 2007. Engaging the Global Countryside: Globalization, Hybridity and the Reconstitution of Rural Place. *Progress in Human Geography* 31 (4): 485–507.

Woods, M. 2012. New Directions in Rural Studies? *Journal of Rural Studies* 28 (1): 1–4.

8 Shaping Public Spaces in Rural Areas

Lessons from Villages in the Gmina of Krobia, Poland

*Karolina Dmochowska-Dudek, Marcin Wójcik,
Paulina Tobiasz-Lis, and Pamela Jeziorska-Biel*

Introduction

The Territorial Context and Characteristics of Domachowo, Stara Krobia and Potarzyca

The work detailed in this chapter encompassed Domachowo, Stara Krobia and Potarzyca – villages situated in the western Polish gmina (local unit of administration) of Krobia that constitute three of the 12 localities making up the folk microregion called *Biskupizna* (of which Krobia is the "capital") (Figure 8.1) *Biskupizna* is located in the south-west of Poland's Wielkopolska region. The work took up the issues of how local communities operate and what influence on shared space they can exert in a period of change associated with the impact of the EU's Common Agricultural Policy. Such matters come under the concept of neo-endogenous development, which has very much coloured the debate on the EU's policy towards rural areas.

From the Middle Ages through Poland's partitions in the 18th century, this land formed part of the Estate of the Bishops of Poznań. Krobia was the locality in which the Bishops residence was located. Thanks to the more-limited burden associated with socage[1] under the Bishop's rule, the peasants' economic situation here was far better than for people governed by Poland's nobility. Throughout the Partitions period (1772–1918), *Biskupizna* was part of the Krobia Holding, managed directly by the Prussian administration. This meant a relatively early and favourable enfranchisement of peasants that helped to encourage certain individualistic and elitist feelings manifesting themselves here, even today, in a lively folk culture that distinguishes this area from other subregions of Wielkopolska. At present, *Biskupian* folk culture is popularised by four folk bands active in the gmina of Krobia. While research on *Biskupian* folk culture is in the scope of ethnology or cultural anthropology (since the end of the 19th century[2]), little has been written about its role in transforming social space.

Due to the historical context (material and non-material heritage), *Biskupian* villages are a great example of transforming social space through designing places as described by Woods (2011) and Wójcik et al. (2019), as well as the relationships in the community, with a focus on rural lifestyles, but also ways in which local people actually experience rural space (Halfacree 2006; Price and Evans 2009; Riley 2010). The three selected villages – Domachowo, Stara Krobia

DOI: 10.4324/9781003110095-10

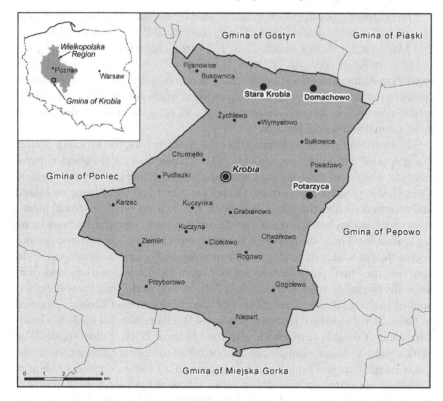

Figure 8.1 Locations of the villages studied within the gmina of Krobia.
Source: Author: Karolina Dmochowska-Dudek.

and Potarzyca – exemplify places where community action shapes public spaces in each village. The broader context of these community-led interventions can be seen within the debates about the rapidly changing demographic landscape of rural regions, reflecting how social diversity can raise economic potential and activity, allowing the renewal of rural values.

Shaping Public Spaces – The Theoretical Context

There are definite properties characteristic of public space, i.e. the space provided for joint activities among community members and other users (such as tourists). Such space is under the partial control of various players in the local community, whose interests it serves, and it is organised, not least in line with a boundary between the public and the private determined quite frequently by relevant institutions. Both the users of the space and their needs are considered important. From their point of view, public space ought to secure safety, aesthetics, accessibility, attraction ("magnetism") and sociability, while continuing to assure that those different activities can be engaged in (Bierwiaczonek and Nawrocki 2012; Bierwiaczonek 2016).

The centre of a Polish village should further be understood as constituting the area of public space defined in Article 2, point 6, of Poland's Spatial Planning and Management Act of March 27th, 2003 (the *Dziennik Ustaw* Official Journal of Laws RP of 2003, no. 80, item 717, with subsequent amendments). This regulation defines the area of public space as one of special significance when it comes to the needs of inhabitants being satisfied, quality of life improved and the establishment of social contacts encouraged – all on account of location, as well as functional and spatial features.

The essence of the work detailed here lies in identifying the various practices being applied within the process of positive change that rural Poland is experiencing, thanks to rural renewal and especially the shaping of public spaces. Overall, this progress results from people's creativity and effort being mobilised and coordinated, also thanks to the support received via various external incentives. Such an understanding of the situation meant empirical findings being set against local rural development theories, especially via the neo-endogenous approach – in which the endogenous part refers to bottom-up development. In contrast, the "neo" part identifies various supra-local manifestations and their local development roles (Ray 2006, 279). In our work, this framework was applied to three villages of the gmina (local-authority area) of Krobia, as located in Poland's Voivodeship (province-region) of Wielkopolskie – a modern administrative unit roughly corresponding to the historical Wielkopolska region. The work sought to target examples of actions taken to improve living conditions, foster integration and bridge social disparities. This research aimed to identify initiatives associated with designing public spaces in rural areas and determines the conditions that need to be met if the practices of shaping public spaces are the drivers for village development.

Sources of Information and Methodology of Research

Our research is based on quantitative and qualitative data – gathered both in the field (the primary data) and desk research (secondary data). The methodology merges elements of in-depth interviews (as augmented by mind maps relating to the perception and evaluation of public space in the villages), focus groups including the most important stakeholders in local development and an investigative walk taken to support the research questions. Such methodological triangulation is beneficial in enhancing the quality of research while at the same time limiting any potential errors of interpretation. In this particular case, the application of various complementary investigation techniques allows for multiple testing of the self-same hypothesis, overcoming limitations and shortcomings of each technique. This triangulation was applied intentionally to deepen the investigation and broaden the phenomenon under study. In this connection, eight in-depth interviews were conducted with local village leaders and activists.

In the village of Domachowo, the in-depth interviews were complemented by mental mapping of how public space in the village is perceived and evaluated. Mental maps of our interviewees helped to exemplify social representations of investigated places. They can be interpreted within the concept of "threefold complexity" after Halfacree (2006), which expresses the contemporary multidimensional understanding of "countryside" and "rurality".

More widely, the integration of space and society has been considered by the idea from Lefebvre (1991) relating to "socially constructed" space. This received further input from D. Massey (1992, 2005) who argued that "the social is spatially constructed too" and therefore "geography matters" as well as from Soja (1999) and Thrift (2000, 2003). This relational understanding of spatiality is focused on interpreting temporal and spatial distribution patterns of social phenomena that manifest themselves in such key spatial qualities as siting, institutional designation, boundary-setting (inclusion and exclusion), scale-setting, location (the neighbourhood effect), distance (proximity vs. isolation), place linking and transformation (spatial change driven by policy). Mental maps of proximate space (meaning the neighbourhood in the circumstances of the countryside) are illustrations of socially constructed, perceived space, full of symbols, significances and values attributed by members of the local community to specific places, and allowing for a deeper interpretation of this dual relationship between communities and their spaces.

The focus groups were established to include the most important stakeholders in the development, albeit representing various sectors and villages. Therefore, the Mayor and Deputy Mayor were invited, along with official stakeholders from the Gmina Office, village leaders, representatives of NGOs and local activists. Use was also made of the aforementioned mobile interactive method of the "investigative walk". This was applied to the three villages, in line with a plan reflecting the research dispositions. Their participants were the local stakeholders directly involved in projects renewing public spaces, i.e. Goat Market in Domachowo, the village community centre and the refurbished area around the grocery shop in Stara Krobia. The walks took the form of a structured narrative on the village. They came to be seen as both an appropriate and vital component of the study.

The collected materials were analysed in line with the Theory of Change (ToC) approach as a tool that stresses clear specification of the logic underpinning any intervention (Connell and Kubisch 1998; Blamey and Mackenzie 2007; Taplin and Clark 2012). The aim is to facilitate subsequent monitoring of a policy process in line with a logical chain of cause-and-effect linkages, with changes set against clearly specified goals to achieve their evaluation. ToC was preferred to other approaches because of its stronger focus on the complex socio-economic and institutional processes underlying social change and the key role in stakeholder participation.

Results

The Rural Public Spaces and Their Perception

Domachowo

Domachowo is a locality in the western part of the gmina of Krobia, some 6 kilometres distant from the centre of Krobia itself (Figure 8.1). This village has an agricultural and service-related character, with a landscape dominated by

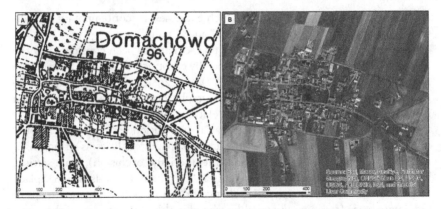

Figure 8.2 The spatial development pattern in Domachowo: A – in 1933, B – in 2018.

Source: A – A detailed map of Poland (1:25,000), by Polish Military Geographical Institute, B – World Imagery by Esri, DigitalGlobe, GeoEye, i-cubed, USDA FSA, USGS, AEX, Getmapping, Aerogrid, IGN, IGP, swisstopo, and the GIS User Community.

typical farmyard structures (Figure 8.2). It is compact, with buildings laid out around an irregular square. Behind the farmyards, two side roads facilitate access to fields and barns. Farm fields are separated by internal roads that mark the boundaries of landholdings and run in parallel. Farmers' success in consolidating farm plots has, at last, ensured an increase in areas of land owned, with the result that there are now fewer tracks between individual fields. Entry routes lead into the village and join together near the church. Overall, the settlement structure in Domachowo should be seen as very consistent with elements of the natural environment. The fields surrounding the village provide a background to it, while roads beyond fences form a borderline between working space and residential areas. Those looking at the village from the outside will see clearly a compact, harmonious silhouette within which the church tower is the main dominant feature. The red roofs of buildings instil a sense of order, cohesion and continuity of development, as strengthened by traditionally placed greenery. Valuable old trees protect the historic central part of the village, and especially its "social heart" – the church. Built of wood, the latter is obviously a place of particular importance to the residents of Domachowo. The Parish Church of St. Michael the Archangel was erected in 1568 under Bishop Adam Konarski. It was rebuilt in 1775 and extended in 1930. The church's historical values and spatial persistence is so significant that it serves as a "cultural anchor" for individual inhabitants as well as for the entire local community. It is a very important and irreplaceable element of Domachowo among the three village centres.

Given this valuable historical and cultural heritage, as well as certain settlement-related features, Domachowo has gained the status of historical settlement (also extending to the spatial setting of this "oval village"). Under these circumstances, cultural assets in general and certain buildings receive protection and care, with items included on the Monuments Register. The settlement's urban (spatial) patterns are important, as are the landscape-related and architectural perspectives.

1. church 2. car park 3. welcome stand 4. cemetery 5. former Inn 6. Goat Market 7. pond and bench 8. community-centre facility

Figure 8.3 Centres of the village of Domachowo.

Source: Authors' own elaboration; World Imagery by Esri, DigitalGlobe, GeoEye, i-cubed, USDA FSA, USGS, AEX, Getmapping, Aerogrid, IGN, IGP, swisstopo, and the GIS User Community.

The local community identifies three centres of Domachowo village fulfilling different functions (Figure 8.3), i.e.:

- Centre 1: around the Church and also including the car park, welcome stand, cemetery, former Inn and Goat Market nearby;
- Centre 2: around the community-centre facility, playground and bandshell;
- Centre 3: around the pond with a bench and a nearby shop.

The common feature here is the serving of social integration, concentration of inhabitants, and provisioning of places to meet. In addition, the first centre is the village's historic heart, important hierarchically, serving a representative function, located at the entrance to the village close to the Church, and playing a significant role in the community. In the opinion of our interviewees, this place is also an all-accessible centre, regional and cultural – owing, as well, to the presence of the former Inn (*Gościniec*) – "the place where the story of the village begins" (PL1PubLI[3]). The latter is a facility serving communal integration and one of the two places where the local community meets up. It plays the role of Cultural Centre of *Biskupizna*, hosting exhibitions, conferences and regular meetings of the Village Housewives Circle. It is also a base for Domachowo's song-and-dance ensemble, while numerous workshops, events and festivities are held here. This is also a hierarchically important place, identified by people in the village's vicinity and socially active inhabitants. Traditional lace making workshops are held here frequently, as are sculpture classes, painting sessions, and events of a culinary or fashion-related profile. The purpose is to promote (to outsiders but also upcoming generations) *Biskupizna* folk culture and its customs such as cooking, baking and crafts making utilitarian objects. The output of these efforts is distributed as the occasion arises, sold or given away as inhabitants of Domachowo share their culture in various ways.

Not far from the former Inn is the Goat Market, a smallish square surrounded by flower beds and equipped with benches where repose and relaxation are possible. The square, referred to as "small or pocket-sized", actually gives rise to greater controversy among inhabitants. By no means everybody likes it, as some consider it a source of spatial dissonance – "it is a good place, but not the best, the reservations refer to potential lack of consistency with the village's character" (PL1PubLI).

The second centre comprises the area around the community local integration facility, which is at the other end of the village from Centre 1. The building as such fails to evoke inhabitants' adoration, instead of encouraging resort to terms, like "communist", "from the socialist period", "grey", "contrasted with the Church". However, these opinions relate solely to image and dubious aesthetics, as the functions discharged here are much appreciated. This is an important, much-needed, functional object and the venue for various activities among children and young people. The playground adjacent to it imparts further order to this space. The bandshell is also located here – as another site for various popular events, including the *Biskupizna* Festival.

The third centre is characterised by a typical location for historical, square-centred villages. It encompasses a pond and a bench (currently being revitalised), as well as a shop. This is another place that naturally attracts attention and positive emotions of residents and newcomers.

Stara Krobia

Stara Krobia is situated in the northern part of the gmina of Krobia (Figure 8.1). Just like Domachowo, this is a centre supporting farmers and also, supplying relevant services. The very first documented mentions of the village are from the 15th century. Still, the settlement existed long before that, serving as the venue for local fairs. This locality is also under conservation protection. The pattern of roads and buildings again underpins this village's morphological classification as oval. Only some of the buildings along the road towards Domachowo stretch out into the form of a traditional "row village" (Figure 8.4). The initial spatial pattern of settlement is well preserved, with new developments making explicit reference to it. While the dominant kind of dwelling is the single-storey gabled structure with ridges running parallel to the road, there are also two-storey buildings with flat roofs. In contrast, new dwellings in the village's western part are single-storey and flat-roofed. The official Protection Registers list numerous residential and farm buildings from the turn of the 20th century, located in this village. The locality of Stara Krobia has no significant industrial assets since the dominant function of this settlement unit is agricultural production (there are 87 farms, and output is based primarily on those that are family-run).

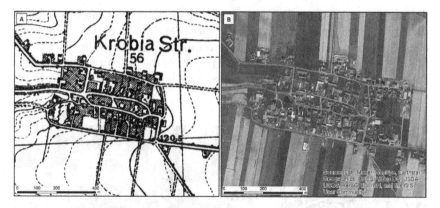

Figure 8.4 The spatial development pattern in Stara Krobia: A – in 1933, B – in 2018.

Source: A – A detailed map of Poland (1:25,000), by Polish Military Geographical Institute, B – World Imagery by Esri, DigitalGlobe, GeoEye, i-cubed, USDA FSA, USGS, AEX, Getmapping, Aerogrid, IGN, IGP, swisstopo, and the GIS User Community.

According to the head of the village, Roman Olejniczak (PL2PubLI), the central places of Stara Krobia are constituted by:

- Centre 1: the area in which the day centre (Świetlica), sporting facilities and playground are all situated,
- Centre 2: the village community centre,
- Centre 3: the area around the pond, with the bonfire site and benches, and
- Centre 4: the refurbished area around the grocery shop (with greenery and benches).

All these places integrate the village's inhabitants. The day centre, erected at Mr Olejniczak's initiative, is the inhabitants' pride and joy since it results from their joint efforts and hard work. These days, it is the venue for the harvest festival and other events through the summer season. The community centre serves different functions as a place for public meetings, dances, culinary workshops, "Catherine" cooking parties and project work involving schoolchildren. The third of the centres is the renovated area around the pond, where young people mostly choose to meet. Older inhabitants prefer to spend time in the summer around the grocery shop, where new greenery and benches have been installed. "In the village, you feel like part of a family" (PL2PubLI).

Potarzyca

The name "Potarzyca" is most probably derived from the expression *po targu*, meaning "after the fair". Potarzyca predates even Krobia as a venue for fairs. It is situated at a crossroads for North–South and East–West routes, hence having really convenient mercantile conditions. Only when a Castellany was established in the more-secure Krobia, fairs moved there instead. In the Middle Ages, the significance of this village was greater than today – a church noted here in 1419 vanished without a trace 100 years later. Most probably, there was a parish here between the 12th and 16th centuries. The villages in the vicinity of Krobia were established in the spatial pattern of the oval form. In the case of Potarzyca, this initial pattern may still be traced, especially at the heart of the village (Figure 8.5).

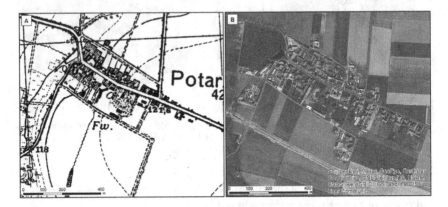

Figure 8.5 The spatial development pattern in Potarzyca: A – in 1933, B – in 2018.

Source: A – A detailed map of Poland (1:25,000), by Polish Military Geographical Institute, B – World Imagery by Esri, DigitalGlobe, GeoEye, i-cubed, USDA FSA, USGS, AEX, Getmapping, Aerogrid, IGN, IGP, swisstopo, and the GIS User Community.

Figure 8.6 Biskupian folk culture in the public space of Domachowo (2018).

Source: Author: Marcin Wójcik.

Potarzyca has just a single centre, namely the green area, around which the village local community facility is located, along with the Fire Station and the kindergarten. The area around these buildings has been restored. It now includes a gazebo, a soccer field and a beach-volleyball field. Also, the greenery has been enriched significantly, and an outdoor exercising area is now in place. The concentration of all public space in just one location ensures a high degree of usage. Thus, the community facility hosts local associations' meetings (including the Circle of Village Housewives), workshops for children and occasional events (such as Seniors' Day, Women's Day, Children's Day, Farewell the Summer and dances). The community facility hall is also used for commercial purposes, with revenue for the village earned in this way (PL4PubLI, PL5PubLI).

Community Initiatives Shaping Rural Public Spaces

In the three selected villages – Domachowo, Stara Krobia and Potarzyca – it is possible to distinguish similar strategies and practices and conditioning of the shaping of public spaces – starting with the planning phase through implementation and then evaluation of the actions undertaken. Also of importance is the actual functioning of the space in question, its perceptions among inhabitants and the later possibilities for it to be maintained and cared for. The objective is to improve inhabitants' living conditions and promote rural areas, attracting tourists and increasing tourist traffic (the village "going outside"). The fact that inhabitants express their feeling of care for the common space and willingness to cooperate in activities undertaken to shape their future design, reveals the raised awareness and sense of responsibility for their local environment that inhabitants feel and manifest. Local initiatives, often constitutive for activities, associated with shaping public spaces in Domachowo, Stara Krobia and Potarzyca have involved various important projects implemented using internal and external assets.

During the period 2012–2016, seven initiatives were implemented in the studied villages. The first project, entitled "Green villages of Biskupizna", was implemented in Stara Krobia in 2012 within the framework of Edition Five of the "Act Locally 2012" grant competition. The action's objective was to enhance awareness and a sense of responsibility among the inhabitants of Stara Krobia, where the environment and ecology are concerned. Project implementation increased the amount of total green spaces in Stara Krobia, with illegal waste dumps attended to, new trees (maples and ashes) and shrubs planted, and walkways put in, near the village's sports facilities. A square was established at the village centre, where a wooden hut was constructed and greenery planted. Workshops were run for inhabitants of the village, with ecological and environmental tasks for the *Biskupizna* Primary School pupils. In parallel, the public space was developed in Potarzyca by creating an aesthetically pleasing and safe place for sport and recreation behind the Farmers' Centre (*Dom Rolnika*), being the main local community centre in the village. The project achieved integration of the village dwellers in common action to ensure a safe and attractive place of recreation for the youngest inhabitants, stimulate further activity and implement successive initiatives. Four years later, the area around the community centre was paved to improve the functionality and aesthetics of the former Inn in Potarzyca, and a basketball court was also developed. The "bug hotel" was installed using plants potentially useful for bees to make honey. This project brought about other benefits from the presence of these useful insects. In this context, a lime tree was also planted, becoming the symbol of common efforts to renew and develop Potarzyca. Additional planting further beautified and diversified the area, augmenting the existing greenery.

The main objectives of the actions implemented in Domachowo were to reinforce the village's attractiveness and allow for the organising of a Festival of Tradition and Folk Culture that promotes the *Biskupizna* folk region. The first project – "Domachowo – a village with tradition" (2013) was focused on welcoming elements symbolising the *Biskupizna* folk culture. An entry gate with sculpted columns was placed in front of the village community centre. The area next to the historic wooden parish church was also refurbished, with a car park and information board installed (the latter telling readers about the locality's history and its main cultural attributes). The task pursued (re) generated places where inhabitants can and do spend their leisure time, play together and strengthen social ties. At the same time, they may transmit to the younger members of local communities a knowledge of Domachowo's history and traditions and organise space for open-air events. Two years later, the project entitled "Domachowo – the capital of *Biskupizna*. Whirling in the dance, tasting tradition" provided the introduction and improvement of street furniture elements, such as benches, bins, roofed picnic tables and bike stands. The community centre exteriors were also painted, its fence replaced, equipment on its playground improved and greenery enriched by new planting, including an orchard's development. A reinforcement of the village's attractiveness was achieved through joint pursuit, with inhabitants, of a concept for developing part of the area at the village community facility. In 2016, the local community organised an exhibition of photographs supplied by village dwellers, and an open-air cinema was also an example of an interesting cultural event integrating local community. The Festival operated via nine-themed stands presenting regional cultural heritage via craft products, musical instruments, herbs, regional publications and regional cuisine.

The Anatomy of the Action

The objective of the aforementioned projects shaping rural public spaces has been to improve inhabitants' living conditions and promote the countryside, with the likely result that tourist traffic increases (as the village "goes outside"). As many inhabitants representing various interest groups in local communities under study care for common space and reflect their interest in cooperating over undertaken activities, this reflects a high level of awareness of, and responsibility for, the local environment. In this context, the action takes the form of shaping the public space of a village via a set of interventions that develop new places or regenerate existing social infrastructure.

Intermediate Outcomes and Causal Pathways

The action forms two causal pathways framed by material and non-material aspects of the transformations brought about. The first focuses on developing the village in a spatial sense; the second on the development of intangible elements – via a bottom-up approach (Figure 8.7).

First, there is a chain of intermediate outcomes, starting with the implementation of projects concerning public space development. As a direct consequence that space assumes greater significance. All of this translates into a higher quality of life (material living conditions in the countryside). The second chain of intermediate outcomes has a strong non-material element. It is about developing human and social capital – as reflected in growing capacities to draw down resources for the local community's benefit. It is also about the development of local-community activity and, consequently, raising the subjective quality of life.

The linking part in the two chains entails becoming more effective in obtaining the European Union funds for so-called "hard" – infrastructural projects transforming rural space. This is possible thanks to institutional organisations shaped by residents themselves, associations possessing legal personality and NGOs. As part of the so-called "soft" EU programmes, funds acquired by Domachowo, Stara Krobia and Potarzyca are dedicated to developing individual skills and strengthening the social capital of these villages. This happens due to, for example, rural communities' increased involvement in both spatial and social initiatives.

Internal Assumptions

While five internal assumptions may be noted, the foremost among those described below should be emphasised because they are starting points for change regarding the villages' spatial development.

A – the starting point for the action is the existing social capital of the villages under study. This clarifies how important is the initial level of social capital in the countryside if any change (including spatial) is about to start. Thanks to the cooperation, trust and established networks, the local community can reach out for the EU funds.

B – a fundamental internal assumption underpinning the success of activity relating to rural public spaces in the area under study concerns the local assets of three *Biskupian* villages. What is involved here are the resources of material and non-material heritage described above.

C – leaders (including the moderator) – the presence of a leader who instils the ideal of renewal in inhabitants and enthuses them into activity in this domain. These are primarily heads of villages, who concentrate the most-active inhabitants around them and encourage their engagement. There should be collaboration and joint work with the Gmina Office, with a distinct emphasis placed on the role of a moderator as a person providing support and participating actively in the process by which public space is shaped. Leaders were needed here as the trigger for the process, but the local community was then ready to work together.

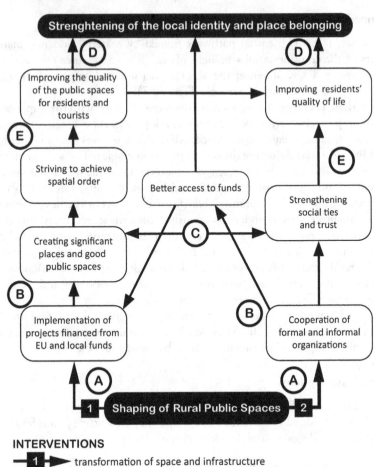

Figure 8.7 Theory of change map for the shaping of action over public spaces in rural areas.

Source: Authors' own elaboration.

D – The next "internal" contingency is associated with the dual connection between demographic stability/growth and the goal of rural renewal in general. This link is contingent upon demographic and economic growth or their sustainability associated with enhanced well-being.

E – Strategic capacity of all Association members in defining, reflecting and pursuing their aims to build up a platform for sociocultural activities in Krobia and the learning capacity of the Association, including its flexibility and adaptability to react effectively when opportunities arise. In each of the villages considered, there is a Local Action Group in operation (the "Together for Old Krobia" Initiating Group, the "Bolder Together" Association for the Development of the Village of Potarzyca and the Group for the Renewal of the Village of Domachowo).

Discussion

In this chapter, we address several key contexts underpinning the creation and shaping of rural public spaces. The first of these relates to local specifics and social change as a feature extending through time. This means that the "heroes" in our case are local communities from a culturally strong microregion of Poland (with a specific folk tradition forming part of the wider cultural features of the Wielkopolska region) – hence a focus of research on the role cultural heritage can or should play in contemporary local development. A second context relates to the level of activity displayed by a local community as well as the mechanism by which public spaces are designed or shaped in contemporary rural areas reflecting rapid processes of functional changes towards multifunctionality (in essence at the expense of agriculture-only) but also, the ageing of rural society, local-authority activity and the founding of rural institutions (signalling the formalisation of activity). A third context then links up with securing the EU funding as a key means of pursuing the development of rural areas in various spheres of local life – especially via the operation of the institutions referred to above and the development of new or improved buildings and areas of public spaces.

In countries like Poland, it is possible to point to distinctively new economic and social outcomes of the EU policy targeting rural areas. Since the share of the population that is rural remains high here (at around 39%), along with the high proportion of the working population still employed in farming (ca. 12%), this leaves a situation in which social and economic issues in state policy towards the countryside are more important than in many Western European states. Irrespective of any assessment of actual programmes seeking to intensify local activity (with a view to community integration being enhanced), Poland's rural communities can be thought to retain cultural capital strong enough to allow for real territorial attachment to shape place-based local development actions (Wójcik 2017).

Further leading questions here evidently relate to the development of sustainable rural communities (Zegar 2012) and the search for development strategies that match the various relevant options (e.g. Rosner and Stanny 2014; Wójcik and Czapiewski 2016). Many researchers feel there is no single solution to how rural areas develop (e.g. Ward and Brown 2009; Naldi et al. 2015). And indeed, this view is more and more reflected in the EU policy's espousal of a "place-based approach", with recognition of the needs of particular rural communities and the involvement of diverse local-development stakeholders (as under "Community-led local development" – CLLD).

Our results confirm how strategies and policies for the development of rural areas need to consider the latter's considerable diversity, particularly basing themselves on place-specific resources. This is even seen to be true at the level of individual villages ostensibly similar to one another but still, in fact, possessing their unique material, social and cultural assets. Thus, the self-same development conditions offer many different models for change in rural communities, as framed both long ago at the time of their genesis and by the current situation. However, there is no single notion or concept to fit each community (e.g. Ward and Brown 2009). This is why the place-based approach needs to relate to lower administrative organisation levels of states and regions, just as much as to regions in their entirety (Naldi et al. 2015).

Indeed, current conditions in the EU Member States (and especially those acceding in the 21st century) allow the concept of neo-endogenous development to serve as an effective model describing the scale and directions of change in rural areas. The work we conducted confirms the achievement of development objectives like bottom-up initiative, multi-partner engagement, place-based orientation, innovation at the level of solutions and the development of good practice, decentralisation of (multi-sourced) funding and the founding of cooperation networks at different levels of relationship, both formal and more informal. Thus, it is possible to note the repeatability of outcomes in different areas impacted by the EU supported instruments (Cejudo and Navarro 2020).

It is further necessary to concur with Terluin (2003), who described the outcome of the neo-endogenous approach for countries that had earlier begun to pursue projects reconciling common European objectives and inhabitants' needs in particular localities. First and foremost, the specific structures and processes and the social and cultural genesis (and hence specific identity) of given rural areas should be noted. At the same time, a factor global in nature was also seen as important, i.e. the social dynamic to change conditioned by ever-greater migratory movement and the consequent establishment of hybrid local communities. This reflects how social diversity can raise the level of innovation – and hence economic potential-and activity allowing rural values to be renewed (Bosworth 2006; Stockdale 2006; Herslund 2011). Administering diversity is, for sure, a difficult process. Still, the result can be synergy to many different players' activity in the life of a community (e.g. Flynn and Marsden 1995), who draw their ideas from different (local, regional and global) tiers of socio-economic development. A huge role is played by various kinds of knowledge and the ability of effective communication between stakeholders in local development processes (Terluin 2003; Ward and Brown 2009; Wójcik et al. 2019). This kind of conceptualisation of the neo-endogenous approach allows for identifying and interpreting phenomena whose sources are in local space (the local community) and national sociocultural frames or at the European Community and its regional level policy (Kockel 2002; Keating et al. 2003; Kasabov 2014). These are then brought together with relational space thanks to liaison via institutions, action groups and stakeholders in change (Ray 2001).

The three Polish villages we studied make clear that the shaping of given places' specifics is today a demanding but much-needed activity that strives to instil a state of social, economic and cultural equilibrium (Ray 2000; Ward et al. 2005; Syssner 2009; Bosworth and Atterton 2012).

Conclusion

We have considered villages with different numbers of central places and diverse ways of utilising them. Potarzyca has just a single such location – the village's local community facility and its surroundings. In Domachowo and Stara Krobia, there are centres identified in several locations. The mayor of the gmina of Krobia Michał Listwoń (PL13PubLI) opined that public space these days is limited to communal (gmina) – run space. The setting and distribution of this space within the village arise out of the land-ownership structure, which is conditioned historically.

Each of the three villages has its own local specifics, even as they have a common quality relating to the need and will to respond adequately to challenges associated with the development of common public spaces. In that context, all three localities have shared traits that constitute the ensemble of preconditions for contemporary social activity. These in particular encompass:

- the compact nature of the settlements, ensuring proximity of spatial and social neighbourhoods;
- the existence of a historical continuum of public spaces in the form of squares within the villages, as well as other forms of the spatial pattern having pro-social properties, road networks and hydrographic objects (ponds);
- the locations of facilities rendering services and/or strengthening economic, social and cultural functions, i.e. churches and other religious buildings, community centres, Fire Stations, old manor houses, inns, schools and other educational facilities;
- the farming character of the localities involved, meaning a definite social commonality of problems and activities.

If public spaces in rural areas are to be successfully well shaped, the several overlapping conditions need to be met:

1. the presence of a leader who instils the idea of renewal among inhabitants and enthuses them into activity in this domain (these are primarily heads of villages, who concentrate the most active inhabitants around them and encourage their involvement in undertakings);
2. openness and a will to collaborate with the local leader in a collaboration that should gain reflection in inhabitants' involvement in both small-scale

activity for the good of the village (e.g. restoring public space to order and organising local events), and strategic undertakings putting local programmes of renewal into effect;

3. the collaboration that should also encompass local organisations, associations and other legal entities, which ought to act in support of renewal, in particular through an active search for sources of financing of the activities undertaken;

4. collaboration and joint work with the Gmina Office, with a distinct emphasis on moderators as people offering support and participating actively in the process of shaping public space.

The above factors have to coexist over a longer period of time. This will facilitate the pursuit of a consistent idea of village development by the local community. A strategy of this kind will then yield satisfaction among inhabitants and concrete material benefits that put in motion a process of self-propelled development. Initiatives shaping public space help to develop social capital in villages. Also, it is possible to note that human capital is converted into social capital (as local leaders' skills and competencies generate enhanced social trust, intensified networks of relations, and shared norms and values – see Putnam 1993, 2000).

An essential effect relating to the rural renewal concept that is observed in Polish villages of Domachowo, Stara Krobia and Potarzyca, being the subject of our study, is how residents themselves have initiated a change in their locality and mounted joint attempts to improve their quality of life and economic independence in this way developing a kind of engine for development. At the same time, inhabitants here feel a sense of responsibility that is translated into what they can do for themselves and for future generations, caring for identity and not foregoing modern technical solutions or concepts available and taking advantage of the other EU Member States. The care taken to ensure the preservation of identity (i.e. elements allowing the separate character of the given locality to be distinguished, recognised or indicated) in this case relates to the traditional folk culture of the *Biskupizna* microregion.

Acknowledgements

This chapter's authors are part of the research project "RELOCAL. Resituating the Local in Cohesion and Territorial Development". The authors acknowledge funding from the European Commission's Horizon 2020 research and innovation programme under Grant Agreement No 727097. For more details, visit the project website www.relocal.eu.

Notes

1 Socage is "(…) form of land tenure in which the tenant lived on his lord's land and in return rendered to the lord a certain agricultural service or money rent" (www.britannica.com).

2 In the late 1870s and the 1880s, Oskar Kolberg, Polish ethnographer and folklorist, conducted field research in Wielkopolska region and published a series of regional monographs which describe the Polish folk culture of the 19th century (see: www. oskarkolberg.pl).
3 This is the code identifying the given respondent.

References

Bierwiaczonek, K. 2016. *Społeczne znaczenie miejskich przestrzeni publicznych*. Katowice: Wydawnictwo Uniwersytetu Śląskiego.

Bierwiaczonek, K., and T. Nawrocki. 2012. Teoretyczne spojrzenie na przestrzeń publiczną. In *Rynki, malle i cmentarze. Przestrzeń publiczna miast śląskich w ujęciu socjologicznym*, ed. K. Bierwiaczonek, B. Lewicka, and T. Nawrocki, 23–64. Kraków: Wydawnictwo Nomos.

Blamey, A., and M. Mackenzie. 2007. Theories of Change and Realistic Evaluation: Peas in a Pod or Apples and Oranges? *Evaluation* 13 (4): 439–455.

Bosworth, G. 2006. Counterurbanisation and Job Creation: Entrepreneurial In-Migration and Rural Economic Development. *Centre for Rural Economy Discussion Paper*, 4: 1–15.

Bosworth, G., and J. Atterton. 2012. Entrepreneurial In-migration and Neoendogenous Rural Development. *Rural Sociology* 77 (2): 254–279.

Cejudo, E., and F. Navarro, eds. 2020. *Neoendogenous Development in European Rural Areas. Results and Lessons*. Cham: Springer.

Connell, J. P., and A. C. Kubisch. 1998. Applying a Theory of Change Approach to the Evaluation of Comprehensive Community Initiatives: Progress, Prospects, and Problems. *New Approaches to Evaluating Community Initiatives* 2: 1–16.

Flynn, A. and T. K. Marsden. 1995. Rural change, regulation and sustainability. *Environment and Planning A*, 27(8): 1180–1193.

Halfacree, K. 2006. Rural Space: Constructing a Three-Fold Architecture. In *Handbook of Rural Studies*, ed. P. Cloke, T. Marsden, and P. H. Mooney, 44–62. Thousand Oaks, CA: Sage.

Herslund, L. 2011. The Rural Creative Class: Counterurbanisation and Entrepreneurship in the Danish Countryside. *Sociologia Ruralis* 52 (2): 235–255.

Kasabov, E. 2014. *Rural Cooperation in Europe: In Search of the Relational Rurals*. London: Palgrave Macmillan.

Keating, M. J., J. Loughlin, and K. Deschouwer. 2003. *Culture, Institutions and Economic Development: A Study of Eight European Regions*. Cheltenham: Edward Elgar.

Kockel, U. 2002. *Regional Culture and Economic Development: Explorations in European Ethnology*. Aldershot: Ashgate.

Lefebvre, H. 1991. *The Production of Space*. Oxford: Blackwell.

Massey, D. 1992. Politics and Space/Time. *New Left Review* 196: 65–84.

Massey, D. 2005. *For Space*. London: Sage.

Naldi, L., P. Nilsson, H. Westlund, and S. Wixe. 2015. What is Smart Rural Development? *Journal of Rural Studies* 40: 90–101.

Oskar Kolberg Institute 2015. *Biography*. Accessed January 10, 2021, http://www.oskarkolberg.pl

Polish Military Geographical Institute 1929–1939. A Detailed Map of Poland 1:25,000. Accessed March 15, 2021. http://maps.mapywig.org

Price, L., and N. Evans. 2009. From Stress to Distress: Conceptualizing the British Family Farming Patriarchal Way of Life. *Journal of Rural Studies* 25 (1): 1–11.

Putnam, R.D. 1993. *Making Democracy Work: Civic Traditions in Modern Italy*. Princeton, New Jersey, USA: Princeton University Press.

Putnam, R.D. 2000. *Bowling Alone: The Collapse and Revival of American Community*. New York: Simon & Schuster.

Ray, C.. 2000. Endogenous Socio-economic Development in the European Union – Issues of Evaluation. *Journal of Rural Studies* 16 (4): 447–458.

Ray, C. 2001 *Culture Economies: A Perspective on Local Rural Development in Europe*. Centre for Rural Economy. https://www.ncl.ac.uk/media/wwwnclacuk/centreforruraleconomy/files/culture-economy.pdf

Ray, C. 2006. Neo-endogenous Rural Development in the EU. In *Handbook of Rural Studies*, ed. P. Cloke, T. Marsden, and P. H. Mooney, 278–290. Thousand Oaks, CA: Sage.

Riley, M.. 2010. Emplacing the Research Encounter: Exploring Farm Life Histories. *Qualitative Inquiry* 16: 651–662.

Rosner, A., and M. Stanny. 2014. *Monitoring Rozwoju Obszarów Wiejskich. Etap I. FEFRWP*. Warszawa: IRWiR.

Editors of Encyclopaedia Encyclopaedia Britannica, 1998 s.v. "Socage". Chicago: Encyclopedia Britannica. Accessed January 10, 2021, https://www.britannica.com/topic/socage

Soja, E. W. 1999. Thirdspace. Expanding the Scope of the Geographical Imagination. In *Human Geography Today*, ed. D. Massey, J. Allen, and P. Sarre, 260–278. Cambridge: Polity Press.

Stockdale, A. 2006. Migration: Pre-requisite for Rural Economic Regeneration?. *Journal of Rural Studies* 22 (3): 354–366.

Syssner, J. 2009. Conceptualizations of Culture and Identity in Regional Policy. *Regional and Federal Studies* 19 (3): 437–458.

Taplin, D., and H. Clark. 2012. *Theory of Change Basics. A Primer*. New York: Actknowledge. https://www.actknowledge.org/PDFs/PACETheoryofChangeDiscussionPaper.pdf

Terluin, I. J. 2003. Differences in Economic Development in Rural Regions of Advanced Countries: An Overview and Critical Analysis of Theories. *Journal of Rural Studies* 19 (3): 327–344.

Thrift, N. 2000. Pandora's Box? Cultural Geographies of Economies. In *The Oxford Handbook of Economic Geography*, ed. G. L. Clark, M. P. Feldman, M. S. Gertler, and K. Williams, 689–704. Oxford: Oxford University Press.

Thrift, N. 2003. Space: The Fundamental Stuff of Geography. In *Key Concepts in Geography*, ed. N. Clifford, S. Holloway, S. P. Rice, and G. Valentine, 85–96. Thousand Oaks, CA: Sage.

Ward, N., and D. L. Brown. 2009. Placing the Rural in Regional Development. *Regional Studies* 43 (10): 1237–1244.

Ward, N., J. Atterton, T. Y. Kim, P. Lowe, J. Phillipson, and N. Thompson. 2005. Universities, the Knowledge Economy and Neo-Endogenous Rural Development. *Centre for Rural Economy Discussion Paper Series* 1: 1–15.

Woods, M. 2011. *Rural*. London, New York: Routledge.

Wójcik, M., ed. 2017. *Tożsamość I miejsce. Budzenie Uśpionego Potencjału Wsi*. Łódź: Wydawnictwo Uniwersytetu Łódzkiego.

Wójcik, M., and K. Czapiewski, eds. 2016. Multifunctional Development in Rural Spaces: Challenges for Policy and Planning. *Studia Obszarów Wiejskich* 43: 1–136.

Wójcik, M., P. Jeziorska-Biel, and K. Czapiewski. 2019. Between Words: A Generational Discussion about Farming Knowledge Sources. *Journal of Rural Studies* 67 (2019): 130–141.

Zegar, J. S. 2012. *Współczesne Wyzwania Rolnictwa*. Warszawa: Wydawnictwo Naukowe PWN.

9 Mother's Little Helper

A Feminist Political Ecology of West Africa's Herbicide Revolution

William G. Moseley and Eliza J. Pessereau

Introduction

"Mother's Little Helper," a hit tune in 1966 by the rock group the Rolling Stones, was a dark and catchy ode to valium. The song's lyrics describe a women's reliance on this increasingly available drug to solve a structural problem of sorts (Wikipedia 2021). Clearly, the mothers of 1966 America were in an impossible situation, shackled by gender norms and patriarchal structures. Under these conditions, valium – made widely available by the pharmaceutical industry – was an understandable response (Waldron 1977). West Africa's female farmers have been placed in an equally impossible situation, attempting to farm their fields with limited power to requisition household labor, a shrinking rural labor pool and increasingly problematic weeds. The emergence of widely accessible and increasingly affordable herbicides has helped them solve this problem for now, but the health consequences of this strategy of using agrochemicals (like valium for housewives in the Rolling Stones' tune) have yet to fully emerge. Of course, male farmers have also adopted the same strategy and have been at it longer than most women. But African female farmers represent, in many ways, the last frontier for this global herbicide revolution, a technological crutch that has arguably short-circuited a realignment of gender norms in the African context as well as the development of safer and more agroecological strategies to manage weed problems. Greater recognition of these health and environmental concerns may lead to change, just as we began to see shifts, albeit inadequate, with gender roles and responsibilities in American households once problematic structures of repression and inequity received greater recognition and attention.

During fieldwork in Southwest Burkina Faso in 2019, we found that 92% of 141 female farmers interviewed are now using herbicides on a routine basis, representing a dramatic increase from previous decades (Gianessi and Williams 2011). In this chapter, we seek to better understand what factors are driving herbicide use by female farmers in Southwestern Burkina Faso. More specifically, we pursue three interrelated questions, operating at different scales, in order to address this broader question. First, how are changes in the global herbicide industry influencing herbicide uptake at the local level? Second, how do economic changes at

DOI: 10.4324/9781003110095-11

the regional level influence the supply of agricultural labor? Third, how does the relative power of women to control labor within their households influence their adoption of herbicides as a labor-saving technology?

Using a feminist political ecology framework, we explain this shift as primarily driven by three factors operating at different scales. First, the rise of generic herbicide production in India and China since the early 2000s means that this labor-saving technology is increasingly affordable in West African markets. Second, artisanal gold production has siphoned labor away from farming systems across West Africa, further constraining female farmers' access to labor. Third, women have limited control over household labor and thus face serious labor constraints in their own farming efforts. While increasing herbicide use is a rational response to labor constraints, it also contributes to growing health risks as well as the spread of herbicide-resistant weeds. The rest of this chapter is organized as follows. First, we will situate this study in the relevant geographical literature. Then we present the study methodology and present our findings organized in terms of the three research questions. We conclude by summarizing our major findings and proposing some policy recommendations.

Context in the Literature

This chapter is informed by three bodies of scholarship dealing with: (1) the political economy of the global agrochemical industry, (2) the New Green Revolution for Africa (GR4A) and (3) women's labor in African farming systems This section reviews each of these themes in turn.

Global Economic Changes and the Agrochemical Industry

Geographers and political ecologists have long been interested in impacts of broader scale political economy on local level human–environment interactions. In the realm of agriculture and related areas such as lawn care in the Global North, the dynamics and shape of the global agrochemical and seed industry has the ability to shape how farmers interact with their landscape. For example, Robbins (2007) convincingly showed how large lawn care companies in the United States (firms that produce the seeds, fertilizers and herbicides used on American lawns) were able to drive the uptake of agrochemicals by middle-class Americans even when these same consumers had health and environmental concerns about using such products. He further discussed how consolidation in the lawn care industry was spurring more aggressive marketing drives. Others have written about a trend of increasing corporate concentration in the global agrifood sector that is influencing farmer behavior (Clapp and Purugganan 2020). Luna and Dowd-Uribe (2020) have written eloquently about the efforts of Monsanto to encourage the uptake of Bt cotton in Burkina Faso. Others, such as Haggblade et al. (2017a) and Clapp (2021), have discussed how the production of generic herbicides in India and China has led to the rise of herbicide use in Africa.

The New Green Revolution for Africa (GR4A)

The conventional wisdom is that the first Green Revolution largely bypassed Africa because it did not focus on African crops (Toenniessen et al. 2008.). The Green Revolution was a concerted attempt to promulgate the use of hybrid seeds, pesticides and fertilizers in the Global South (Moseley 2015). Since the mid-2000s, there has been a second phase of this Green Revolution, known as the GR4A, that has targeted African smallholders. The architects of the GR4A have sought to improve food production and nutrition by encouraging small farmers in Africa to purchase improved seeds and inputs and increasingly sell their crops at a market (Moseley, Schnurr and Bezner Kerr 2016). While the GR4A is similar to the first Green Revolution, namely in its focus on improved seeds, pesticides and fertilizers, it is also different in important ways. While the first Green Revolution was largely state-led, the GR4A has embraced the role of the private sector and often celebrates public–private partnerships. The GR4A also sees market integration as critical, often conceptualizing this in terms of value chains. Lastly, the GR4A has emphasized gender and the role of women in African farming systems (Gengenbach et al. 2018). While the GR4A has stolen much of the limelight in African agricultural development circles in recent years, we would argue that there are many shifts happening in global agricultural input and commodity markets that overshadow the GR4A as a force for African agrarian change (as discussed above). As such, we suggest that the GR4A may be a ripple in a larger shifting sea and that agricultural geographers working in Africa should pay closer attention to some of these broader dynamics and aspects of globalization (a major theme of this book). It is not that the GR4A is not contributing to some of these trends, but it may be the lesser driver amidst larger forces.

Women's Labor in African Farming Systems

By some estimates, women produce 70% of the food crops in the African context (NEPAD 2013). While some scholars dispute this figure, women are undoubtedly central players in African agriculture (Christiaensen 2017). Despite their prominent role in African food systems, women struggle to secure access to the land, labor and inputs needed for agricultural production. Carney (1993) was one of the pioneering geographers to document struggles over women's labor in farming systems in Africa, specifically in The Gambia. Shroeder (1998) also wrote about the power of women in Gambian households and the role that income from gardening could play in improving their relative power in household decision-making. While there is an emerging literature on the role that herbicides have played as a labor-saving technology in African farming systems (Sheahan and Barrett 2017; Luna 2020a), less has been written about the use of herbicides by female African farmers and the factors that may be driving the use or non-use of these chemicals.

Methods

We use political ecology and feminist political ecology as our theoretical frameworks in this chapter. Political ecology examines local human–environment interactions nested within the context of national, regional and global political economy (Robbins 2012; Moseley et al. 2013). In this instance, we seek to understand how local and regional rural labor dynamics, as well as changes in the global herbicide industry, may be influencing local herbicide uptake (and political ecology, with its multiscalar analytical framework, is quite helpful in this regard). Feminist political ecology acknowledges that gender roles and responsibilities within households are socially constructed but these norms have real impacts on livelihood outcomes (Rocheleau et al. 1997). Given the highly gendered nature of work and household responsibilities in rural Burkina Faso, this lens seems especially appropriate for better understanding labor constraints faced by women in their farming activities.

Burkina Faso is an especially appropriate place for this study as research by other authors suggests that herbicide uptake in Africa is most advanced in many of the countries where cotton production is highly developed (Haggblade et al. 2017a; Luna 2020a). Burkina Faso frequently ranks among the top three cotton producers in Africa and cotton production is most advanced in the southwestern part of the country, site of this study (Moseley and Gray 2008; Luna 2020b). We lastly note that women are heavily involved in agriculture in Burkina Faso, both on their own plots and as laborers on the plots of family members.

Fieldwork for this study was undertaken in the June–July 2019 period. Semi-structured interviews were undertaken with 141 female rice farmers in five different communities in Southwestern Burkina Faso (Figure 9.1). This project builds on a longer-term research project dealing with GR4A rice initiatives in the same villages (Gengenbach et al. 2018). This project conducted interviews with these same women in 2016, 2017 and 2019. Background information on each of the women, including that on household structure and wealth, was collected in 2016 and is used in this analysis. Women were originally selected for interviews based on a census the research project conducted of those involved and not involved with a GR4A rice project (and then randomly chosen as a subset of these women farmers).

Women farmers were interviewed in their native language for approximately 45 minutes. They were asked a series of questions about their experience using herbicides, how they apply them, their reasons for using them and the gender dynamics in their household regarding farm labor. We also interviewed 12 herbicide vendors at the village level and three herbicide wholesalers plus one importer in Bobo Dioulasso, a city of a half million people and the major commercial hub for Southwestern Burkina Faso. These vendors and wholesalers were asked about marketplace dynamics and changes in sales over time. They were selected for interviews based on who we found on the side of the road when we visited a village and, for the wholesalers and importer, based on the advice of staff working for the GR4A rice project. Our positionality, as white North Americans (male and female), undoubtedly had some influence on the way farmers

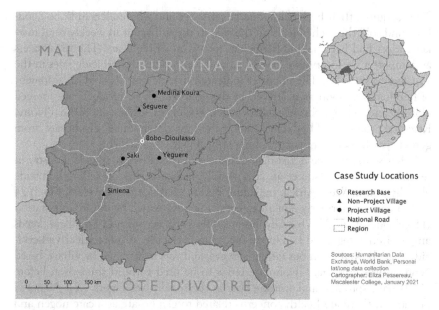

Figure 9.1 Map of field sites in Southwestern Burkina Faso.

Sources: Humanitarian Data Exchange, World Bank, Personal lat/long data collection. Cartographer: Eliza Pessereau, Macalester College, January 2021.

and vendors responded to questions. We were quite careful in explaining that we were not associated with the GR4A rice project, but some interviewees may have still assumed this to be the case.

Findings

In this section, we present our findings related to a broad examination of the factors driving herbicide use by female farmers in Southwestern Burkina Faso. In the first subsection, we explore how changes in the global herbicide industry are influencing herbicide uptake at the local level. In the second subsection, we examine how economic changes at the regional level influence the supply of agricultural labor. In the third subsection, we explore the dynamics at the local level, with a particular focus on labor constraints and gender dynamics for female farmers.

Changes in the Global Herbicide Industry and Supply Chain to Burkina Faso

The majority of women we interviewed indicated that the increasing availability and affordability of herbicides played a big role in their decision to adopt herbicides. As noted previously, we interviewed 12 herbicide vendors at the village level and three herbicide wholesalers plus one importer in Bobo Dioulasso. All of the village-based vendors also worked as farmers as their primary occupation.

They acquired their herbicides from larger shops and wholesalers in Bobo Dioulasso and sold these chemicals on the side in the village or in weekly rural markets. Most had been selling herbicides for two to three years. The wholesalers and importer we interviewed noted a significant uptick in herbicide sales in the past five to six years. The importer indicated that he quickly sold his shipment of herbicides as soon as it arrived in the country. These products are sourced globally and in neighboring countries such as Guinea, Ghana and Côte d'Ivoire. Regional producers of herbicides import their main ingredient, glyphosate, from generic producers in India and China (Haggblade et al. 2017b).

Other scholars have noted major changes in the global herbicide industry in recent years (Haggblade et al. 2017a). Glyphosate is the top-selling herbicide globally and in West Africa, including Burkina Faso. It was developed in 1974 by Monsanto, a formerly American company now owned by Bayer Corporation (Haggblade et al. 2017b). Glyphosate kills both grasses and broadleaf weeds and is often categorized as a broad-spectrum herbicide. A more selective herbicide that showed up in our farmer surveys, and was mentioned by village-based vendors, was atrazine. Glyphosate tends to be used early in the season before plants have sprouted and atrazine later in the season (see Figure 9.2). As noted previously, there are health concerns related to glyphosate as a carcinogen and hormone disruptor (Mesnage et al. 2015), and for atrazine as a teratogen and hormone disruptor (Hayes et al. 2002).

Since 1994, regional rules and regulations have been streamlined to facilitate the importation of pesticides, including herbicides, into Sahelian West African states. In that year, a regional body known as the *Comité Permanent Inter-Etats*

Figure 9.2 Atrazine (left) and glyphosate (right) are commonly used herbicides in Burkina Faso.

Source: Photos by William G. Moseley.

de Lutte contre la Sécheresse dans le Sahel, or CILSS, established a regulatory committee to review and certify pesticides for importation into member countries
(Diarra 2015). If pesticides pass efficacy and safety tests, then they are cleared
for importation into all nine CILSS member states, making this an attractive
market for pesticide manufactures (Haggblade et al. 2017b). The concern, of
course, is that companies target the country with the least stringent health and
safety standards, and then get approval to market their product in all nine CILSS
states.

Developed in 1974, glyphosate was sold exclusively under Monsanto's trademark name *Roundup* until it went off patent in 2000 (Duke and Powles 2008).
After it went off patent, it began to be produced by a number of other companies,
at first those based in the Global North (Bayer, DuPont, Syngenta, BASF, Crop
Science and Dow) and then in the Global South. Sales of the herbicide grew
from US$5.46 billion in 2012 to US$8.79 billion in 2019 (Richmond 2018).
Sales took off in tropical countries after Global South producers (mainly India
and China) offered even lower price points for the herbicide. In Africa, rates of
adoption have been uneven but especially high in the cotton production areas
of West Africa (Grabowski and Jayne 2016). In India, herbicide use tripled
between 2005/2006 and 2015/2016 (Das Gupta et al. 2017). In Mali, Haggblade
et al. (2017b) found that glyphosate herbicide prices had declined 35% in the
local CFA currency (the same regional currency used in Burkina Faso), or 50%
in US dollars, between 2008 and 2015. Given the price sensitivity of poor farmers, these cost declines are a very significant factor influencing the uptake of
herbicides in the region, and these cost declines are directly related to changes
in the global herbicide industry.

Rural Labor Dynamics and Artisanal Gold Mining

While not a major part of the empirical fieldwork undertaken in this project,
an understanding of shifts in artisanal gold mining in the region sheds light
on increasingly constrained rural labor dynamics. Gold mining in the region is
not a new activity but stretches back over a millennium (Kevane 2015). While
gold mining was at the center of a series of African kingdoms (most notably
the Ghana, Mali and the Songhay Empires) that ruled in West Africa from the
ninth through the fifteenth centuries, large-scale industrial mining returned to
the area in the 1980s and is a leading source of income for Burkina Faso, Mali
and Ghana (Moseley 2014). Burkina Faso is the fifth leading producer of gold
in Africa today (Brugger and Zanetti 2020). Alongside industrial gold mining,
which employs few local people, small-scale artisanal mining has also returned.
These small-scale mining efforts have accelerated in the past 15 years due to
a raft of affordable technologies, many of which are produced in China, such
as metal detectors, excavators, crushers and sifters that make the process much
more efficient (Werthmann 2017).

Low-cost technology combined with higher gold prices has spurred the exo-
dus of young people from Burkina Faso's rural areas (Werthmann 2009). Young
people, including young men and women, leave the village to not only earn

money but to have an income that is not controlled by their father or male head of household. Brugger and Zanetti (2020) estimate that in some areas of Burkina Faso, two out of every three rural households have at least one member working in the artisanal gold mining sector. A 2016 Government of Burkina Faso report suggests that 1 to 1.2 million people are directly or indirectly involved in the industry, of a total Burkina population of 18.6 million in that year (Assemblée Nationale 2016). While some more successful miners may invest their profits in farming, buying capital goods like tractors, in many cases there is a tension between farming and mining (Brugger and Zanetti (2020). Pokorny et al. (2019) find that artisanal gold mining has been a major drain on agricultural labor in northern Burkina Faso. This finding is consistent with the overall tight rural labor market we encountered in our study area. While few women brought up gold mining unprompted in our interviews as the cause of labor shortages, although some did, it helps one understand why it is so challenging to hire farm labor of any sort. This, in turn, helps explain the appeal of herbicides as a labor-saving technology.

Household Level Labor and Gender Dynamics

The findings in this subsection are primarily based on semi-structured interviews with female farmers undertaken in the June–July period of 2019. We found that 92% of women in our sample used herbicides in all or some of their own fields (n = 141). This figure is much higher than others have reported for women farmers in the region in the past (Gianessi and Williams 2011). While the majority of women used herbicides in all of their fields, 5% of women only sprayed fields with more remunerative crops, such as rice or maize rather than peanuts or cowpeas, due to lack of sufficient funds. For women who did not use herbicides, the primary reason was lack of funds to pay for herbicides, although a few shared concerns about herbicide effectiveness or health risks associated with herbicide spraying.

On average, women began using herbicides in 2011 or 2012 (seven to eight years prior to interviews). Many women began using herbicides in fields they farmed as soon as these chemicals became available at the village level at a reasonable price. Women noted that previously herbicides could only be found at weekly rural markets; however, now they are sold by village-based vendors (see Figure 9.3). Other reasons women began using herbicides (as cited in interviews) included: (1) acquiring a field to farm,[1] (2) increases in earnings which allowed them to purchase herbicides, (3) increases in the number of weeds in their fields and (4) loss of strength necessary for hand weeding due to old age. While 48% of women (n = 130) apply herbicides on their fields twice per season, other women apply herbicides anywhere from one to seven times per season. Spraying typically occurs before plowing, after planting and after crops have begun growing for several weeks. Although herbicides kill most weeds, 89% of women said that there are some resistant weeds that remain even after spraying, and 33% said that weeds have become more resistant over time, suggesting that herbicide effectiveness has decreased (n = 36).

Figure 9.3 Herbicide seller at the village level (left) and male relative of female farmer applying herbicides in her rice field (right).

Source: Photos by William G. Moseley.

The majority of women do not use paid labor for hand weeding, typically doing weeding themselves with the help of children or other family members. Only 14% of women use paid labor for hand weeding because they suggested that in most cases it is more time and cost-effective to apply herbicides instead. Women typically have a man apply herbicides in their fields, be it a son, husband, other male relatives or paid labor (see Figure 9.3). Of women using herbicides on their fields, 13% employed men to spray herbicides for them (usually when they could not call on a male relative to do this for free). Other studies suggest that herbicide use, even if someone must be paid to apply the chemicals, is more cost-effective than hiring people to do hand weeding. For example, Haggblade et al. (2017b) found that herbicides cost 50% less on average than hiring weeding labor at the farm level in neighboring southern Mali.

In our own analysis, we found that herbicide products plus paid labor, for those who use it, only cost 4% more, on average, than hiring hand-weeding labor. The costs were 21,197 CFA (38.38 USD) per hectare versus 20,380 CFA (36.90 USD) per hectare, respectively. While herbicides can be applied by a single individual, weeding is often performed by a group of 3–4 people. In addition, hand weeding takes up to 60 times longer than applying herbicides ($n = 117$): on average, a single person will take over 33 8-hour days to weed one hectare (the equivalent of 266.7 hours), whereas one person can, on average, spray one hectare with herbicides in 4.5 hours. This time difference, coupled with a general scarcity of labor, helps us understand why most women opt for herbicide use over hiring hand-weeding labor.

The time savings associated with herbicide use, both in terms of a woman's own labor, as well as in the reduction in time it takes to complete the task, is very significant. In Burkina Faso, women in rural areas traditionally spend the majority of their days working in their husband's fields (Kevane and Gray 1999), although there is evidence that some rural women are beginning to rebel against this tradition (Luna 2020a). Outside of these obligations, women often have limited time to tend to their own fields, in addition to foraging for wild edible plants, cooking for the family and caring for children. Women's own fields are particularly important because they provide an important source of income, as well as sauce ingredients, filling the traditional expectation that women in households are responsible for providing the nutrient-dense sauce components (such as peanuts, cowpeas, okra, shea and foraged plants) and men the staple grains (Morgan and Moseley 2020). Women also use earned income to cover school fees and pay for medicine and clothing for their children. Because women have such limited time to work in their own fields, any time and labor-saving techniques have the potential to significantly impact their day-to-day lives, income and ultimately the nutritional variety and food security of the family.

Gender dynamics play a large role in the application of herbicides, as 92% of women do not apply herbicides themselves. Sixty-two (62) percent of the people applying herbicides for women are men: husbands, sons, nephews, brothers-in-law or paid labor. This is a significant shift from hand weeding, where 97% of women either hand weed themselves or do so with the help of family members or paid labor. Therefore, using herbicides not only decreases the amount of time women must work to remove weeds from their fields, in most cases it completely removes the need for them to be involved in any weed removal, thus freeing up more time.

When asked why they have men apply herbicides rather than do it themselves or asking female family members, most women seemed surprised and amused at the possibility of a woman applying herbicides. Many of the responses given were tied to gender roles (e.g., "this is how it is"). When asked to specify, women said that they are not strong enough to use the herbicide pump, whether due to old age or simply because they are women or that men in the house do not allow them to spray. In many instances, women claimed that they simply did not know how to apply herbicides. This could be due to the fact that, in the early days of herbicide use in West Africa, agricultural extension agents invited only men to trainings (Moseley and Gray 2008). In recent years, the spread of herbicides in Burkina Faso has meant that more people are able to learn from neighbors or parents who already have the skills for applying them. Indeed, the women in our sample who apply their own herbicides learned almost exclusively from neighbors or family members, and an herbicide vendor we spoke with said he had taught many women how to use herbicide pumps. One woman even taught herself. Many women stated that if they or their daughters were taught how to apply herbicides, they would gladly do the application themselves. However, it should be noted that women may have felt pressured to answer this way due to the nature of our previous questions. Among more vague reasons aligned with gender roles, 10% of women cited health risks, whether for themselves or for infants they nurse or carry on their backs throughout the day. These concerns are not unfounded, as there are a number of documented health risks associated with herbicide use, including cancer and endocrine disruption (Toe et al. 2013; Mesnage et al. 2015; Bationo 2017).

Several women noted how the herbicide backpack sprayers often leak onto their backs, and they are afraid that their contact with the chemicals would get their infants sick. One woman who applies herbicides herself said that she set her child down while spraying. Even when not actively carrying their children, women worry that spraying herbicides could contaminate their breastmilk and impact their children through nursing. Few women knew where information about these health risks had come from, although some cited television programs or advertisements, and one said that men had told her of the risks. The majority of women who noted health risks could not name a source. An herbicide vendor stated that he was aware of the risks and protective gear that should be worn when spraying (gloves, boots, face mask), but he did not emphasize this to his customers because the gear could not be found locally. It is possible, of course, that he preferred not to warn his customers of health risks for fear of losing business.

The 4% of women who regularly apply herbicides themselves do so because no family members will help them, and they do not have the money to pay someone else (a man) to do it. An additional 11% of women know how to apply herbicides and do so when their families are not able to help them. Because these groups are so small, we were unable to perform statistical analyses to understand if these groups differ from women who do not apply herbicides themselves.

The type of person who helps a woman apply herbicides to her field may be influenced by the age of the woman (Table 9.1). While at least 50% of women in each age group receive help from a male family member, be it a son, husband, brother-in-law or nephew, middle-aged (31–50 years) and older (51–80 years) women receive this type of help more often than young women (18–30 years). On the one hand, this is likely because the majority of respondents have sons who apply their herbicides, and young women may not have sons who are old enough to operate a herbicide pump. Young women, on the other hand, more often apply herbicides by themselves or with occasional help from family, possibly due to the same lack, or lack of control of, household labor. Older women, whose children have grown up and now apply herbicides on their own fields, leaving them little time to do this for their mothers, are more likely to resort to paid labor. Table 9.1 shows the percentages of categories of people who typically help women apply herbicides in their fields ($n = 115$).

In polygamous households, first wives are traditionally considered to have more power than subsequent wives (Akresh et al. 2016). As with age, we wondered if first wives may use more unpaid labor to apply herbicides than subsequent wives, given that they may have more control over household labor and may not need to look elsewhere for help. However, we found that first wives and subsequent wives do not differ greatly in their ability to requisition household labor

Table 9.1 Herbicide application help by women's age

Adult women	All family members (%)	Female family member (%)	Herself (%)	Herself or family member (%)	Male family member (%)	Paid labor (%)
18–30	10.0	3.3	10.0	16.7	50.0	10.0
31–50	15.9	4.8	6.3	9.5	55.6	9.5
51–80	9.5	0.0	0.0	0.0	66.7	23.8

for herbicide application. In terms of paid labor, we see the same result: 16% of first wives (n = 37) use paid labor to apply herbicides, and 16% of subsequent wives (second, third or fourth wives) (n = 45) also use paid labor. Although this is surprising, the relative power of first wives in polygamous households may be diminishing with the more recent phenomenon of household fracturing documented since the 1980s (Marchal 1987).

Conclusion and Policy Recommendations

Female farmers in Southwestern Burkina Faso, despite their relatively limited means, have adopted herbicides with gusto. Women's modest ability to commandeer labor within their households means that their farming activities are labor constrained. Some women also indicate that weeds are a bigger problem than in the past, which could be a result of declining soil fertility and/or greater herbicide resistance among some weed populations. Herbicides offer a way for women to save time and keep up weeding. We found that 92% of the female farmers in our study were using herbicides, about twice per season on average (a rate of adoption and saturation that is higher than past studies have found). Women's limited ability to access labor within the household is further constrained by overall labor constraints exacerbated by migration to cities and artisanal gold mining opportunities in the region.

Women farmers' adoption of herbicides is facilitated by easier access to these chemicals at the village level and at price points that make them competitive with hiring labor to do manual weeding. In all five villages where we conducted surveys, herbicides were easily found with two to three informal vendors per village. The onset of generic production of herbicides in the global South is a large reason for the lower cost of herbicides at the village level, coupled with a regional system that facilitates the importation of herbicides. Some scholars have argued that pesticides cycle through the global marketplace in a predictable pattern. At first, these are sold on patent at a relatively high price point. As they go off patent, and as they become less effective because of pesticide resistance, production shifts to generic producers who offer a lower price point to poorer farmers in areas of the world where pests and weeds have yet to develop a resistance to the pesticide (Haggblade et al. 2017a). Glyphosate seems to be following this pattern, with Africa being one of the last markets for the chemical before it becomes completely obsolete. This chapter's gendered perspective shows how this same cycle continues within communities, as male farmers often have greater means to adopt these chemicals initially, and then as prices decline further, these are more widely adopted by female farmers.

While the upsurge of women farmers' use of herbicides is economically rationale, and helps resolve the nettlesome problem of labor constraints in African farming systems further exacerbated by women's limited power to commandeer household labor, this upsurge in herbicide use does present public health concerns. Previous studies have shown that there are a number of documented health risks associated with herbicide use, including cancer, harm to unborn children and endocrine disruption (Toe et al. 2013; Mesnage et al. 2015; Bationo 2017).

A number of policy responses should be considered by African governments, regional bodies (like CILSS) as well as international organizations such as the WHO and FAO. First, while the CILSS regional system facilitates the certification of pesticides for use in member states, this also means that pesticide producers may focus their efforts on one member state and obfuscate health risks in order to break into this regional market. A more rigorous vetting of these chemicals is needed in order to mitigate health risks. Second, it is also important to recognize the very real labor constraints faced by women farmers. Developing alternative, low-cost approaches to weed control, and privileging women farmers in extension efforts, is critical. There are, for example, a number of agroecological methods that could be further developed to tackle weed problems and address labor constraints.

West Africa's herbicide revolution among women farmers is a relatively quiet revolution and certainly one that is operating under the radar as compared to the much louder, more public and more media-savvy GR4A. The irony is that this quiet herbicide revolution is bringing about more profound changes to African farming systems than its louder GR4A counterpart. However, accompanying this quiet revolution are very real health concerns, like those associated with "Mother's Little Helper" in 1960s America, concerns that deserve much more overt acknowledgment and responses than they are currently receiving.

Acknowledgments

This work was supported by grants from the Womadix Fund and the National Science Foundation (NSF), grant #1539833. The other co-PIs in the NSF group, for a project examining value chains associated with the GR4A, include Rachel Schurman (University of Minnesota), Thomas Bassett (University of Illinois), William Munro (Illinois Wesleyan University) and Heidi Gengenbach (University of Massachusetts Boston). We also wish to express our appreciation to five Burkinabe research assistants who participated in fieldwork for this project: Melanie Ouedraogo, Eveline Héma, Yacouba Zi and Salimata Traore and Bureima Kalaga. Last but not least, we wish to thank our official collaborator, the Burkina Faso National Institute for the Environment and Agricultural Research (INERA), and especially agroeconomist Adema Ouedrago for helping us identify research sites and coordinate household surveys.

Note

1 Most women do not have secure customary land tenure rights (aka usufruct rights) but must work through a male relative to get access to land (Kevane and Gray 1999). The one exception to this is rice fields in some areas where women hold the customary rights and pass them down from mother to daughter.

References

Akresh, R., Chen, J. J. and Moore, C. T. 2016. Altruism, Cooperation, and Efficiency: Agricultural Production in Polygynous Households. *Economic Development & Cultural Change*. 64 (4): 661–696.

Assemblée Nationale du Burkina Faso. 2016. Rapport General de La Commission d'Enquête Parlementaire sur La Gestion Des Titres Miniers et la Responsabilité Sociale des Entreprises Minières. Ouagadougou.

Bationo, H. 2017. Usage incontrôlé des pesticides et impact sur l'environnement et la santé des populations. *Le Faso.* Jan 15.

Brugger, F. and J. Zanetti. 2020. "In my village, everyone uses the tractor": Gold Mining, Agriculture and Social Transformation in Rural Burkina Faso. *The Extractive Industries and Society.* 7 (3): 940–953.

Carney, J. 1993. Converting the Wetlands, Engendering the Environment: The Intersection of Gender with Agrarian Change in the Gambia. *Economic Geography.* 69 (4): 329–348.

Christiaensen, L. 2017. Agriculture in Africa – Telling Myths from Facts: A Synthesis. *Food Policy.* 67: 1–11.

Clapp, J. 2021. Explaining Growing Glyphosate Use: The Political Economy of Herbicide-Dependent Agriculture. *Global Environmental Change.* 67: 102239.

Clapp. J. and J. Purugganan. 2020. Contextualizing Corporate Control in the Agrifood and Extractive Sectors. *Globalizations.* 17 (7): 1265–1275.

Das Gupta, S., B. Minten, N. C. Rao and T. Reardon. 2017. The Rapid Diffusion of Herbicides in Farming in India: Patterns, Determinants, and Effects on Labor Productivity. *European Journal of Development Research.* 29: 596–613.

Diarra, A. 2015. *Revue des politiques sur les pesticides et les produits vétérinaires dans l'espace CEDEAO.* Laboratoire d'innovation FSP – Document de Travail N° West Africa-JSR-2015-2. East Lansing: Michigan State University.

Duke, S. O. and S. B. Powles. 2008. Glyphosate: A Once-in-a-Century Herbicide: Mini-Review. *Pest Management Science.* 64 (4): 319–25.

Grabowski, P. and T. S. Jayne. 2016. Analyzing trends in herbicide use in Sub-Saharan Africa. *Michigan State University International Development Working Paper.* 142: 1–25.

Gengenbach, H., R. Schurman, T. Bassett, W. Munro and W. Moseley. 2018. Limits of the New Green Revolution for Africa: Reconceptualising Gendered Agricultural Value Chains. *The Geographical Journal.* 184 (2): 208–214.

Gianessi, L. and A. Williams. 2011. Overlooking the Obvious: The Opportunity for Herbicides in Africa. *Outlooks on Pest Management.* 22 (5): 211–215.

Haggblade, S., B. Minten, C. Pray, T. Reardon and D. Zilberman. 2017a. The Herbicide Revolution in Developing Countries: Patterns, Causes, and Implications. *European Journal of Development Research.* 29: 533–559.

Haggblade, S., M. Smale, A. Kergna, V. Theriault and A. Assima. 2017b. Causes and Consequences of Increasing Herbicide Use in Mali. *European Journal of Development Research.* 29: 648–674.

Hayes, T. B., A. Collins, M. Lee, M. Mendoza, N. Noriega, A. A. Stuart and A. Vonk. 2002. Hermaphroditic, Demasculinized Frogs after Exposure to the Herbicide Atrazine at Low Ecologically Relevant Doses. *Proceedings of the National Academy of Sciences.* 99 (8): 5476–5480.

Kevane, M. 2015. Gold Mining and Economic and Social Change in West Africa. In *The Oxford Handbook of Africa and Economics: Volume 2: Policies and Practices,* ed. C. Monga and J. Y. Lin. Oxford: Oxford University Press.

Kevane, M. and L. C. Gray. 1999. A Woman's Field Is Made at Night: Gendered Land Rights and Norms in Burkina Faso. *Feminist Economics.* 5 (3): 1–26.

Luna, J. K. 2020a. Pesticides Are Our Children Now: Cultural Change and the Technological Treadmill in the Burkina Faso Cotton Sector. *Agriculture and Human Values.* 37: 449–462.

Luna, J. K. 2020b. Peasant Essentialism in GMO Debates: Bt Cotton in Burkina Faso. *Journal of Agrarian Change.* 20: 579–597.

Luna, J. K. and B. Dowd-Uribe. 2020. Knowledge Politics and the Bt Cotton Success Narrative in Burkina Faso. *World Development.* 136 : 105127 https://doi.org/10.1016/j.worlddev.2020.105127

Marchal, J. 1987. En Afrique des savanes, le fractionnement des unites d'exploitation rurales ou le chacun pour soi. *Cahiers des Sciences Humaines.* 23 (3–4): 445–454.

Mesnage, R., N. Defarge, J. Spiroux de Vendômois and G. E. Séralini. 2015. Potential Toxic Effects of Glyphosate and Its Commercial Formulations below Regulatory Limits. *Food and Chemical Toxicology.* 84: 133–153.

Morgan, J. and W. G. Moseley 2020. The Secret Is in the Sauce: Foraged Food and Dietary Diversity among Female Farmers in Southwestern Burkina Faso. *Canadian Journal of Development Studies.* 41 (2): 296–313.

Moseley, W. G. 2014. Artisanal Gold Mining's Curse on West African Farming. *Al Jazeera English.* July 9. http://www.aljazeera.com/indepth/opinion/2014/07/artisanal-gold-mining-west-afric-20147372739374988.html

Moseley, W. G. 2015. Food Security and the Green Revolution. In *International Encyclopedia of Social and Behavioral Sciences,* ed. J. Wright, 307–310. Amsterdam: Elsevier.

Moseley, W. G. and L. C. Gray. 2008. *Hanging by a Thread: Cotton, Globalization and Poverty in Africa.* Athens: Ohio University Press and Nordic Africa Press.

Moseley, W. G., E. Perramond, H. Hapke and P. Laris. 2013. *An Introduction to Human-Environment Geography: Local Dynamics and Global Processes.* Hoboken, NJ: Wiley/Blackwell.

Moseley, W. G., M. Schnurr and R. Bezner Kerr. 2016. *Africa's Green Revolution: Critical Perspectives on New Agricultural Technologies and Systems.* Oxford: Taylor & Francis.

NEPAD. 2013. *African Agriculture, Transformation and Outlook.* New Partnership for African Development (NEPAD), November 2013, 72 p. Johannesburg, South Africa. https://www.un.org/en/africa/osaa/pdf/pubs/2013africanagricultures.pdf

Pokorny, B. C., von Lübke, S. D. Dayamba and H. Dickow. 2019. All the Gold for Nothing? Impacts of Mining on Rural Livelihoods in Northern Burkina Faso. *World Development.* 119: 23–39.

Richmond, M. E. 2018. Glyphosate: A Review of Its Global Use, Environmental Impact, and Potential Health Effects on Humans and Other Species. *Journal of Environmental Studies and Sciences.* 8: 416–434.

Robbins, R. 2007. *Lawn People: How Grasses, Weeds, and Chemicals Make Us Who We Are.* Philadelphia: Temple University Press.

Robbins, P. 2012. *Political Ecology: A Critical Introduction.* 2 nd edition. Malden, MA: Wiley-Blackwell.

Rocheleau, D. E., B. Thomas-Slayter and E. Wangari. 1997. *Feminist Political Ecology: Global Perspectives and Local Experience.* New York: Routledge.

Sheahan, M. and C. B. Barrett. 2017. Ten Striking Facts about Agricultural Input Use in Sub-Saharan Africa. *Food Policy.* 67: 12–25.

Shroeder, R. A. 1998. *Shady Practices Agroforestry and Gender Politics in The Gambia.* Berkeley: University of California Press.

Toe, A. M., M. Ouedraogo, R. Ouedraogo, S. Ilboudo and P. I. Guissou. 2013. Pilot Study on Agricultural Pesticide Poisoning in Burkina Faso. *Interdisciplinary Toxicology.* 6: 185–191.

Toenniessen, G., A. Adesina and J. DeVries. 2008. Building an Alliance for a Green Revolution in Africa. *Annals of the New York Academy of Sciences.* 1136 (1): 233–242.

Waldron, I. 1977. Increased Prescribing of Valium, Librium, and Other Drugs—An Example of the Influence of Economic and Social Factors on the Practice of Medicine. *International Journal of Health Services*. 7 (1): 37–62.

Werthmann, K. 2009. Working in a Boom-Town: Female Perspectives on Gold-Mining in Burkina Faso. *Resources Policy*. 34: 18–23.

Werthmann, K. 2017. The Drawbacks of Privatization: Artisanal Gold Mining in Burkina Faso 1986–2016. *Resources Policy*. 52: 418–426.

Wikipedia. 2021. Mother's Little Helper. Accessed June 8, 2021 (https://en.wikipedia.org/wiki/Mother%27s_Little_Helper).

Part III

Rural Innovations and Urban–Rural Partnerships

10 Towards a Strategic Model for Sustainable Agriculture in Mediterranean Countries

A Case Study of the *Cooperativa Hortec* (Catalonia, Spain)

Joan Tort-Donada and Jordi Fumadó-Llambrich

Introduction

We undertake a detailed examination of some of the specific implications of the current processes of globalization and transformation of rural economies in Catalonia, a region in the western Mediterranean that occupies a relatively small area (32,000 km², 6.3% of Spanish territory) but which, despite this, has a varied agricultural mosaic – one that is especially representative of what is often referred to as the Mediterranean farming model – and an agrarian economy with a long historical tradition. Indeed, it was this tradition that ceded ground in the economic structure to the industrial sector, a shift that Catalan society experienced from the end of the 18th century (Vilar 1964).

At the core of our examination lies an analytical case study of an agricultural cooperative, the *Cooperativa Hortec*, an initiative that, despite its humble origins, today leads the way in the agro-ecological subsector in Catalonia. The roots of this cooperative – a veritable agricultural company – are comparable to those of many traditional agricultural holdings in Catalonia (i.e. small- to medium-sized, family concerns with high productive diversification); yet, it showed, from the outset, a degree of creativity and a capacity for innovation within the sector – both at the product level and in its commercial practices – that have established it, with the passing years, as a point of reference in the sector. Founded in 1991 by a small group of farmers, the association oriented its activity exclusively – and in a pioneering way – to the growth of organic produce. Adopting a cooperative business structure – a structure that boasts a long tradition in Catalonia – *Hortec* explicitly defends innovative values such as agro-environmental sustainability and the adoption of an integrated production and marketing line – designed to "connect" small producers throughout Catalonia with centralized distribution centres, including Barcelona's fruit and vegetable market, *Mercabarna*. In our discussion, we aim to contribute to the debate on innovation and entrepreneurship in the rural world raised in this book.

It is in this specific context, and by means of this case study, that we seek to address the following fundamental question: In the framework of the present-day agrarian sector in Catalonia (and, more broadly, in the coastal territories of the Western Mediterranean), can an initiative implemented by a small

DOI: 10.4324/9781003110095-13

group of innovative individuals, with local roots in the sector, make a significant contribution to the progressive transformation of that sector in terms of its sustainability without this constituting a threat to the economic, social and family viability of the region's farms? We complement this main discussion by a search for answers to a number of secondary questions or issues that are indirectly linked to our principal question: Is it possible today to speak of an agrarian sector – one operating at different scales – and of a rural world with any real future? In territories with a diversified economy, and where industry and services are prevalent, can agrarian activity interact efficiently with these other sectors? And, in a globalized world do local entrepreneurs/rural dwellers/youth have any real chance of building a future in the primary sector?

In terms of the methodology adopted, we conduct our discussion, which occupies the main body of this chapter, at two levels or scales. The first is more general in nature and seeks to place in their historical context the importance of Catalonia's agricultural sector and the significance that agriculture and the associated *genre de vie* have had on the construction of this region – which we can consider as being representative both of Mediterranean Iberia and, at a larger scale, of Mediterranean Europe. In so doing, we take into account especially pertinent contributions from both geographers and historians as well as the writings of a Catalan writer, Josep Pla, who has explored this subject in depth throughout his work. At the second scale of our discussion, we undertake, within the broader frame of the Catalan agrarian sector, a study of the specific experiences of the *Cooperativa Hortec*, that is, we address its original *raison d'être* and its unique process of foundation and its subsequent development. This process is, as we show, highly illustrative, in many different ways, of the shift undergone by the Catalan agrarian model over the last century, as regards both the decline of the traditional productive model and the expectations awoken within the sector as a result of its gradual insertion in the *global* market and the emergence of a link to products of proximity and organic production. This study is reported in Section A Case Study of the Cooperativa Hortec: Its Foundation, Growth and Present Importance and includes an extensive interview, conducted in May 2019 and June 2020, with two of the founders of the *Cooperativa Hortec*, who, thanks to their strong family ties with the agrarian sector and their active role in the creation of the cooperative, are excellent interlocutors for the analysis we undertake.

Theoretical Insights

The starting point of our discussion can be traced to the specific context of the *intersection of scales* (that is, the global vs. the specific) and of the *complexity* in the relations between space and societies (to the extent that it brings into play highly diverse human realities both as regards their current configuration and their historical development). This, in turn, leads us to highlight a matter that, at the theoretical level, is of some importance – namely, regardless of our decision to focus on an innovative initiative within the agro-ecological sector with a notable projection in its field. The issue we address is also relevant, first, as a manifestation of *geographic intelligence*, understood as a "particular response to given spaces and imperfect societies" (Pueyo-Campos 2020, 778), and, second, in terms of an "integrated approach", within Geography, which enables us to rec-ognize the complexity of the interactions that take place at any given moment in a specific point or *place* in the world, regardless of its size or spatial significance (Murphy 2018, 26).

We also wish to draw attention to the spatial-temporal context in which the case study we address here unfolds. The *Cooperativa Hortec*, which, since it was created in 1991, has been a success story of entrepreneurship and innovation in the Catalan agrarian sector, cannot be analysed in any global sense without tak-ing into account the decline suffered by the rural way of life in European societies over the last century and a half, a process that many authors have borne testi-mony to in works that frequently transcend the local or regional framework and which have acquired a general significance; the case, for example, of Friesland in the Netherlands, as described by Geert Mak (2000), of the peasant farmers of Savoy, in France, as portrayed in John Berger's trilogy (1991), and the Catalan rural world as depicted by Josep Pla (Pla 1952, 1975; Tort and Paül 2008; Tort and Català 2018).

At a more concrete level of analysis, the case of the *Cooperativa Hortec* needs to be framed within the coordinates of the development over the last two dec-ades of new agri-food networks that serve to foster production, distribution and consumption and which are generally conceived as alternatives to dominant agri-food systems (Paül and Haslam McKenzie 2010). However, it also shares a direct link with so-called "short food supply chains", understood as a way of shaping geographic relations of proximity between production and demand (Yacamán et al. 2019, 1), without this necessarily negating the need for "long supply chains" (Morgan 2010, 1861) or implying a high degree of vulnerability (Brown and Miller 2008). The *Cooperativa Hortec* should also be analysed from the perspective of the multifunctionality of rural spaces, where multifunctional-ity is understood not as a simple pretext to justify productivist policies (Woods 2005), but rather in terms of the possibility of connecting this idea with the "multifunctionality of landscapes" (Hernández et al. 2018, 43). The develop-ment of such a perspective is especially interesting in the context of the unique complexity of the traditional agrarian systems of the Mediterranean countries. Finally, we also note that *Hortec's* experience provides an interesting example for discussion in the critical debate on agrarian post-productivism (Tort 1998).

The Context: Catalonia as a Paradigm of the Problems Currently Faced by the Agrarian Sector in the Mediterranean Countries of Europe

Notes on the Historical and Current Relevance of the Agrarian Culture in Catalonia

It has often been noted in relation to various contexts, but especially from the perspective of rural geography, that Catalonia can, for different reasons, be considered a "paradigmatic region" that encapsulates, at a range of different scales – those that we might identify as "Iberian" and "Mediterranean", but also, at the European scale – the basic concept of the so-called *agrarian mosaic*. This diversity, moreover, cannot be separated from a process of human occupation and cultural transformation that was initiated with the protohistoric colourizations and which, with varying degrees of intensity, has been maintained over time. The process is eloquently captured by Arturo Soria (1989, 31) when pointing out that, in Europe, with the exception of its geological structure, all the elements that make up the socio-territorial system can be interpreted in terms of the historical evolution and permanent interaction between the natural environment and human societies.

Within this general framework, the case of Catalonia attracts attention because of the radical nature of its physical characteristics. We are dealing with one of the regions of southern Europe whose mountainous terrain is predominant over the rest of the territory and which constitutes a limiting factor of its agrarian activity (Deffontaines 1975). In a similar line, the historian Pierre Vilar insists on the restrictive qualities of Catalan land and soils, and concludes that the recurrent insufficiency of the cereal harvest to meet the needs of the population, along with the preponderance of poor scrubland and the exceptional nature of land suitable for intensive agriculture meant that "modern" Catalonia – that is, the country that began to take shape at the start of the 18th century – had to opt for an exchange economy: "The general insufficiency of the main food products, the surplus production of raw materials to be transformed or exported, the balancing act between intensive farming settlements and poor areas and the urban markets: everything urged internal or external exchange" (Vilar 1964, 422).

This general vision of the agrarian sector in Catalonia, which to some degree might be extrapolated to other southern European territories adjacent to the Mediterranean with similar characteristics, would be incomplete were we not to consider, along with these physical determinants, the factors of a human and social order that, indivisible from the environment, have served to shape this sector through successive historical periods (Vilà 1973). And here, as we go back in time to the agrarian colonization that took place under Roman rule, it is inevitable that we introduce the concept of the *pagus*. In Roman law, the concept had an important structural legal meaning (the *pagus* defined a whole system of relations between the land and its owner, which included the right to "live" on it and the right to "cultivate" it, as well as the whole regime of relations between the owner and other associates). The *pagus*, and all that has derived from it (both as a legal concept but also its economic and cultural significance), has left its mark on countries once under ancient Roman rule – countries that, today, are known as the Romance countries and which largely coincide with southern and south-western Europe (Tort 2006).

In Catalonia, it is a critical concept for explaining the agrarian tradition of the territory, the predominance of small rural properties – from as early as the Middle Ages – and the historical and economic relevance of the farmer, peasant or *pagès* (a word whose etymology is rooted in the term *pagus*, according to Coromines

1980). Also, to underline the importance of such concepts as "casa" [house] (or *mas* or *masia* [farmhouse]) and "família" [family] to explain the continuity and persistence of a veritable "system of work" and a "relationship with the land" that have been passed down from generation to generation and which, in many respects (tangible, but also intangible), have survived to the present day. The historian, Jaume Vicens-Vives, describes it as follows:

> House and family, *mas* and land: here lie the mighty foundations of the Catalan social structure, (...) especially from the 14th century down to the present day. (...) This fusion between home and family has been promoted by the bond that ties man with the land in the various colonizations of the country.
>
> (Vicens-Vives 1962, 32–33)

This idea of the Catalan farmhouse, the *mas* or *masia*, as a historical representation of the concept of the Roman "pagus" and, at the same time, as a symbol of the small traditional farm of rural Catalonia, is neatly captured, we believe in the landscape shown in Figure 10.1.

Josep Pla, a writer who defined himself, above all, as a *pagès* (as a result of his having been born into a family with a long tradition of working the land), dedicated much of his work (in the form of essays) to the Catalan rural world and the study of its farmers. In the following paragraphs, which clearly capture the ideas expressed by Vicens-Vives, we present some of his more insightful reflections.

Figure 10.1 Image of a traditional Catalan farmhouse, or *masia*, in the eastern Pyrenees.
Source: J. Tort-Donada. November 2017.

Some Thoughts on the Role Played by the Pagès or Catalan Farmer, by Josep Pla

Josep Pla, above and beyond his role as an intellectual (and, arguably, we are speaking of the most important Catalan writer of the 20th century), defended his identity as a *pagès* for two main reasons: on the one hand, on account of his family's ancestry and, on the other, on account of what he himself recognized as a "personal" and "ideological" affinity with the way of life and the way of thinking of the *pagesos* (Tort and Paül 2008, Tort and Català 2018). A short paragraph, at the beginning of the essay that he dedicated specifically to the subject (Pla 1952, 1975), informs us first hand and in a highly succinct fashion of his reasoning on the matter:

> Deep, deep down, I am no more than a *pagès*, pure and simple: a country yokel sophisticated by the culture of our times. (…) I have the impression of being the first owner of my *masia* who has not physically worked the land. (…) Yet, despite my long absences from the land, I have never totally lost contact with it.
>
> (Pla 1975, 10)

Below we present, as a sequence (each contextualized by way of a brief introductory statement), two ideas or thoughts of Pla concerning the meaning and importance of the figure of the *pagès* in Catalonia, not solely from a historical perspective, but also considering the mark they have had, in a broad sense, on Catalan society today.

The Writer's Defence of the Innovative, Versatile and Permanently Adaptive Figure of the Pagès

> The vast majority of our farms have turned their hand to polyculture. They do a little bit of everything, harvesting all the things man needs to survive. The *pagès* or farmer must be highly knowledgeable in many fields. But it is rare that he has the same degree of mastery in each. One farmer will know more about tending vineyards than olive groves; another will be more at home in his orchard than breeding cattle; while the farmer over the way will understand more about pigs than he does about cows or sheep. But no matter where their expertise lies, farmers have to be knowledgeable. It's this diversity that can give farmers a certain aura of wisdom. It's as though they always have something up their sleeve, that they have never revealed everything. In the face of the growing specialization of modern man, this diversity can have a strange effect – exasperating even. The day will come when it will be hard to imagine. The course of modern life tends to guide man along a sole path, as if he were wearing a muzzle.
>
> (Pla 1975, 97)

On the Figure of the Pagès as the "Cornerstone" of Catalan Society

Talking about the farmers, probing a little into their quite peculiar way of being or thinking makes me feel as if I'm actually back in the country. When all is said and done, the number of families that can actually be considered to have severed their ties with their farming ancestry is really quite small. Only a few decades ago important Catalan settlements were towns and villages that lived off the land. Industrialization is barely a century old. The grandparents of the more than half a million people that live in Barcelona cultivated the land. This undoubted ancestry affects the country's way of being. It is clear that in the towns most affected by modern life this basic feature remains more or less apparent. It may vary from one individual to another. But the characteristic always resurfaces, with considerable strength, at all levels of society: among the aristocracy, in the liberal professions, among those that dedicate themselves to industry or trade.

(Pla 1975, 93)

A Case Study of the *Cooperativa Hortec*: Its Foundation, Growth and Present Importance

The ideas and beliefs of the authors cited above – in particular those of Vilar (1964), Vicens-Vives (1962) and Pla (1975) – allow us to infer that, in some senses, the "agrarian substratum" continues to play an important role as a basic element of the present-day collective memory of Catalan society. And all this, regardless of the secondary role played today by agriculture in the economy of Barcelona and Catalonia. This role, moreover, extends well beyond the deep shift – from a rural to urban and metropolitan model of life – undergone by this society since the mid-19th century with the onset of industrialization, the effects of which are still being felt. And this, projected onto our particular case study, allows us to frame the unique nature of *Hortec* within the context of the agricultural sector of the Barcelona region: that is, a symbol and a highly significant example of what we have described above as the *pagès* layer and which, as we have been at pains to point out, has played a role of paramount importance in the agrarian and economic history of Catalonia.

But the potential of an initiative, such as *Hortec*, is not limited by what it represents for present-day Catalonia, be it at the historical or symbolic level. The cooperative is also representative of the opportunities that currently exist in the society and environment in which it operates, within the agro-ecological sector and specifically in the fresh fruit and vegetable subsector. The viability of the initiative, and its establishment in less than two decades as a leader in its field, bring to the fore the values defended from the very outset by the cooperative's founders – values that are now widely shared throughout society: *sustainability*, in terms of agricultural production and distribution), *proximity*, understood, above all, as an "ideal" to be achieved, and the gradual creation of an *agri-food network* between producer, retail distributor and final consumer, that is progressively becoming more and more consistent. The cooperative operates throughout Catalonia but focuses its work above all in the metropolitan region of Barcelona.

The case of the *Cooperativa Hortec* lends itself particularly well to this study for two distinct, yet complementary, reasons. On the one hand, it is an excellent example, within Catalonia's agrarian sector over the last 15 years, of an initiative undertaken in the production and distribution of ecological and local – 0 km – agricultural products. Moreover, it has the territory of Catalonia in general as its productive area of reference and the territory of Barcelona and its metropolitan area (representing 14% of Catalan territory and accounting for 65% of its population) as its "market" (or area of destination for its produce). On the other hand, it provides an outstanding example of the efficacy –albeit an example that is of secondary importance – within the structure of Catalonia's agricultural sector today of a productive layer of first order, yet a layer that is much less visible than the networks of the large agri-industrial and agri-food companies. In practice, the geographical area in which *Hortec* emerged as an initiative, El Maresme, located some 40 km to the northeast of Barcelona, despite its predominantly peri-urban character, is still representative, albeit residually, of the traditional landscape of market garden (or *huerta* in Spanish) production in Catalonia – as shown in Figure 10.2.

It is also worth stressing, at this point, that the initiative to set up *Hortec*, as an "association of *pagesos*" interested in distributing their own production in a more rational and efficient manner, involved adopting the legal form of a cooperative enterprise. That is a type of business organization that has, as its founding philosophy, the promotion of cooperation among its members – rather than having as its primary objective the maximizing of profits for its owners and shareholders, as is more typical of a capitalist firm. Cooperativism,

Figure 10.2 Mixed agricultural landscape in El Maresme, in the north-east corner of the Barcelona metropolitan region.

Source: J. Tort-Donada. March 2018.

as a movement, spread throughout Europe after the end of the 18th century and, in some respects, it developed in parallel with the emergence of the trade unions. In Catalonia, the first cooperatives emerged in the mid-19th century, as production cooperatives, and they enjoyed a great boom, in parallel with the development of the labour movement, between the end of that century and the first decades of the 20th century. It should be noted that, in Catalonia, the cooperatives were especially strong in the territories with an agrarian and rural vocation, but they fell into decline after the Civil War (1936–1939) along with the general slump suffered by the rural world and the abandonment of many of the family farms after 1960.

To obtain an idea of the creation and mission of *Hortec* we conducted an interview with two of its founders, Josep Maria Gamisans and Salomó Torres, who, some 30 years after launching their business, remain linked to the cooperative and are able to offer many insights into its operations.

First, we would like you to provide us with a sketch of Hortec as it is today. Tell us about the type of company it is and the particular characteristics that it has acquired within the agro-ecological sector of Catalonia and Spain, a sector in which we know that over the last 15 years it has established itself as one of its leading firms.

"*Hortec* currently employs 30 people and has an annual turnover of between 10 and 12 million euros. The business itself – i.e., the cooperative – comprises a total of 26 members, that is, 26 farmers, located throughout Catalonia (half of which specialise in horticulture, the other half in fruit growing), who supply the cooperative with most of their production. But our management and distribution network, which adheres first and foremost to the criterion of proximity, has been gradually extended to include external producers – especially when, owing to the nature of the product or weather conditions, demand cannot be met in Catalonia. Regardless of this though, we always ensure that all the conditions and protocols of organic farming are respected. Today, the 26 members supply approximately 30% of all the products we handle and the remaining 70% comes from suppliers in the rest of Spain – a number that must have reached around 300 in recent years. A part of the produce we sell also comes from various European countries. As for our current network of "customers" (that is, retailers or *botiguers*), they range in number between 800 and 1,000. Most of them (80%) are located in Barcelona and its immediate area, 15% are scattered throughout Catalonia and the remaining 5% are located in different parts of Spain."

"The geographical distribution of the farms of the 26 members making up the cooperative is worth mentioning as it constitutes a fairly homogeneous distribution as regards the various regions or "agricultural landscapes" of Catalonia: thus, they are to be found in the coastal areas close to the sea, the case of the farms of Baix Llobregat, Penedès and Anoia, and within the metropolitan region of Barcelona, as well as on the great plain of inland Catalonia, around Lleida – the fruit region *par excellence* – and in the lands to the south, around the river Ebro, characterised by its own special microclimate, and finally in the lands to the north, Catalonia's wet, mountainous territory: the Plana de Vic, Cerdanya, Garrotxa and the region of Girona. It could be said that the diversity of Catalonia is richly reflected in the agro-farming diversity of Hortec's members."

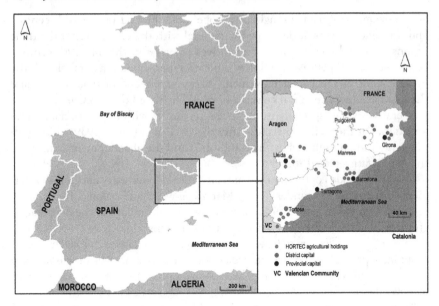

Figure 10.3 Location of the Catalan agricultural holdings directly linked to *Hortec* (2020).
© Wikipedia, Open Database License, ODbL.

The Cooperativa Hortec, founded with the idea of distributing fresh agricultural produce and of bringing farmers and consumers together, was unusual in that it anticipated, by almost two decades, the movement that today campaigns for the sustainability of agrarian systems and local –0 km – agriculture. What we would like to know is the extent to which the cooperative might be considered the outcome of careful, detailed planning or whether, more simply, it is the result of a concatenation of unforeseen factors and personal circumstances.

"The *Cooperativa Hortec* was set up by its eight founding members in 1991. As of today, of those eight, only two of us are still involved [Josep Maria Gamisans and Salomó Torres]. Back in the day, to get a cooperative up and running, you needed to have at least eight members. Many people were interested in setting up a cooperative because it meant, first and foremost, various tax advantages and paying less taxes. But we, and our six colleagues, were more interested in the social side of the business: above all, in a shared project."

"When setting up the cooperative, we were lucky because we were able to do so in association with a chain of stores which, since the early 80s, had been selling organic produce. In fact, *Hortec's* initial core of customers was made up of these stores. So, it was not as if we would have entered into a prior theoretical debate about ecological methods of production; quite simply, we ended up working with them because we had a guaranteed point of distribution for our product."

The fact that the founders and original promoters of Hortec were farmers and that, as such, you had direct ties with the land and agricultural production suggests that this facilitated certain aspects of your development as a company. We would like to know how far you think this was important to your success and to what extent you believe that other less predictable factors were relevant.

"An important role in the growth and expansion of our cooperative can be attributed to a number of circumstantial factors. One such factor was bovine spongiform encephalopathy (BSE) or mad cow disease, which reached Spanish shores at the end of the year 2000. This ushered in a growing demand for ecological products. What makes a Catalan citizen want to eat differently? Well, when he or she starts asking their grocer, "Hey, instead of selling these conventional lettuces, why don't you stock some organic produce?" And the shopkeeper, whose primary goal is to sell and avoid losing a good customer, wastes no time in finding a farmer who produces organic lettuces. And tomatoes, and potatoes, and wine... And it was against this backdrop that *Hortec* emerged. The perfect initiative, at that precise point in time, to satisfy the needs of that particular shopkeeper."

"The reason why this extremely modest initiative should today lead the field in Catalonia's agro-ecological sector is down to its emergence at precisely the right moment in time. By this, we mean the moment when the retail distributor, the shopkeeper, began to be aware of organic produce and began to promote it without any qualms. At the same time, a growing social interest in ecological production was making itself manifest. And this coincided with a situation in which our cooperative was expanding and was financially sound. This meant we were able to go out and look for the produce that interested us, pay for it within the terms requested, not run up any debt and incorporate a number of new partners... Everything just fell into place and began to run like clockwork. In fact, for six or seven years we were the only company in Barcelona distributing fresh organic produce. The boom in demand for organic produce found us in the best position imaginable. We just happened to be in the right place at the right time."

In Catalonia, the cooperative model, especially in the agricultural sector, enjoys a long tradition. From this perspective, Hortec can be seen to fit perfectly within this tradition, while representing a "modern" version of the first cooperative movement – linked also to the first trade unions. To what extent does the legal entity adopted by the company respond to a wish to return to this "classic" concept of the cooperative?

"Truth be told it is impossible to say just how far the fact that Hortec operates as a cooperative has helped us in this whole process of launching and consolidating our business. In Catalonia there is, without any question, a long tradition of agricultural cooperativism, and *Hortec* has, to some extent, participated in that. Yet, it is also true that many myths have been built up around the cooperative movement. Time puts things in their place; in the long run, it will become clear just who has committed themselves to the cooperative system and who has not. Founding a cooperative is not an end in itself. In order for it to work, once created, the farmers who set it up need to remain committed to it. Clearly, the model has not always worked the same way in all parts. We can't speak now of the boom enjoyed by the cooperative movement, a hundred years ago, in the 1920s and 1930s, as that was another era. But some links can be made with the boom recorded in the 1960s, which coincided with the new demand for meat products. And which turned Lleida and Tarragona into leading centres of the cooperative movement."

Figure 10.4 Hortec logo.

Source: Hortec SCCL (Barcelona, Spain).

To conclude, we would like you to take stock of what Hortec has achieved so far and to give us your opinion of the possibilities of undertaking similar initiatives in the sector in the future.

"In retrospect, organizing the business as a cooperative has worked out well for us. However, we are not living in a bubble: in the future, like everywhere else, the uncertainties the sector faces present us with a highly complex scenario. A scenario in which, whatever way you look at it, finding the right direction for a collective project is never easy. Quite simply because, although you might be in quite a strong position, you can never control all the other factors, that is, all the other actors present in the sector. The only thing you can depend on is your own ability to innovate and invest, and your own will to forge ahead. As we said, in retrospect, if *Hortec* is what it is today – a formula for success, it is because we began precisely at a time when the agro-ecological market started to grow and, so, we were able to take advantage of the conditions to manage our growth in an orderly and coherent fashion. All we did as its founders at that time was to believe in ourselves. We were never aware of the significance of what we were doing. Just as those who promoted the first agricultural associations and cooperatives at the end of the 19th century were not aware of the significance of their work."

"What we can't tell you though is what has been more important: setting up as a cooperative or the values that *Hortec* represents and defends. Probably one thing can't be separated from the other."

Discussion

On the "Innovative" Agro-Productive Model of the Cooperativa Hortec and Its Current and Future Viability

In our presentation of the case study we have stressed the importance for the creation and development of *Hortec* of the innovative attitude of its founders and of their pioneering defence of a set of values that, subsequently, have become points of reference for the sector: *sustainability, proximity* and the conception of the company as part of an *agri-food network*. Given the success that the company has enjoyed over the last 15 years, it would seem that an initiative of this kind might be a good model to adopt and one that might be widely "replicated", given an apparent lack of obstacles.

However, we need to be careful not to oversimplify matters. It is more than evident that the foundation of the company was based on a set of values that, insofar as its founders were pioneers in putting them into practice, bestows upon the company a high degree of merit. In addition, both its founding members and those who today constitute the cooperative are *pagesos*: that is, professionals in the sector who, moreover, form part of a historical tradition that continues to maintain a degree of significance in Catalonia today. But it is important to recognize that, in a number of respects, the viability and success of this initiative has depended on a set of "atypical" circumstances – or at least circumstances endowed with characteristics that are unlikely to be reproduced in other contexts. Most notably, the fact of being able to exploit the "market advantage" afforded by the great contrast in population between the metropolitan region of Barcelona (where 80% of the purchasers of the production managed by *Hortec* are located) and the rest of Catalonia (in contrast, demographically very weak, but where most of the farms run by the cooperative members are located together with most of its other suppliers).

Another aspect that we consider fundamental, and one that emerges quite clearly from the interview, is that of the *complexity* inherent in managing an initiative of this kind; a complexity that makes itself most manifest when it comes to integrating, in a synchronized and efficient fashion, the multiple actors involved in the whole process: from producers – the farmers or *pagesos* – and marketers to sellers, trainers and technical and managerial staff, etc. And, here, we should stress the profound significance that the figure of the *pagès* (understood, as we have sought to convey in Section The Context: Catalonia as a Paradigm of the Problems Currently Faced by the Agrarian Sector in the Mediterranean Countries of Europe, as a "global expert" in all matters related to farming and working the land) has had historically, and the need for this significance to be explained to young people interested in entering this sector. Especially because the last two generations of Catalan society have experienced such a marked distancing, or "disconnection", from the realities of the agricultural and rural world, and the primary sector in general, that it seems unlikely that such a process might be rapidly reversed. In fact, this is a problem that we intuit can be extended, *mutatis mutandis*, to a large part of the urban societies of today's world.

On the Historical Significance of Cooperativa Hortec: A Question to Reflect on

An analysis of the territorial growth of *Hortec*, from the company's first dealings until the present day, is highly revealing. That is, both as regards its 26 members and many of the company's current 300 suppliers, the "hard core" of professionals who provide the firm with produce respond, if only to a minimal degree, to the idea of the traditional Catalan *pagès* or farmer, which we have described throughout this chapter. But we are not just talking about a "profession". We are talking about a person who, by vocation, has very strong ties with the land. Ties, moreover, can be very demanding. The diversity of knowledge required to work the land, especially in the context of Mediterranean polyculture, places the figure of the *pagès* at the antipode of specialization and monoculture – the recurrent paradigms of our age. But, here, in our case, there is an added geographical circumstance. Being a farmer or *pagès* in Catalonia, as in most Mediterranean countries in southern Europe, also means assuming the intrinsic diversity of the country's physical environment, with all that this implies in terms of having to be constantly adapting their farming methods. It is no surprise that, faced by this situation, our interviewees should note that "the diversity of Catalonia is richly reflected in the agro-farming diversity of *Hortec's* members".

On the Significance of Cooperativa Hortec for Future Generations of Catalan Farmers

It is also important to stress, we believe, the exemplary value of certain aspects of *Hortec's* "philosophy" for future generations: Cooperation, as the driving force, proved critical in implementing this initiative – and in this sector it is, perhaps, all the more striking, given the individualism that has generally been attributed to the farming classes. But our interviewees – the surviving witnesses of the cooperative's founding group – also point out that "many myths have been built up around the cooperative movement". In a context of many uncertainties, such as those the sector must currently face, it is significant that they consider that "finding the right direction for a collective project is never easy". Moreover, it is clear that their experience has taught them that the future should be addressed with a restrained, prudent and possibilistic attitude. This is to say, in the sense of avoiding the risks of an "excess of optimism", which might be derived from an assessment of the initiative that only takes into account its commercial profitability and economic success.

Finally, there is one point that we believe we cannot ignore in the preceding discussion, especially in the context of the debate on innovation and entrepreneurship in the rural world that is raised in this book. We refer to the permanently open and proactive attitude that characterizes our two interviewees, an attitude that in many ways encapsulates the singular profile – at least in the Iberian and Mediterranean territories – attributed traditionally to the figure of the *pagès*. A profile that is clearly captured in one of their final comments (which defines, in essence, their "plan of action"):

"Although you might be in quite a strong position, you can never control all the other factors, that is, all the other actors present in the sector. The only thing you can depend on is your own ability to innovate and invest, and your own will to forge ahead".

Conclusion

To what extent can the initiative that the eight founders of the *Cooperativa Hortec* launched in 1991 – promoting local, ecological produce, based on principles of sustainability and self-sufficiency – be considered a "strategic model" for those working in this sector, indeed for all those that seek to achieve more equitable and economically viable ways in which to organize the production and commercialization of fruit and market garden products? How might the links between initiatives of this nature and the local areas and regions in which they become embedded be strengthened and protected over time? In the future, just how far can we go in maintaining a relationship of interdependence and recip-rocal correlation between a traditional agrarian sector (as exemplified by Cata-lonia) and the collectives that constitute that sector, as exemplified here in this article by the *pagesos* of the *Cooperativa Hortec*?

Beside conventional agricultural systems, it seems that there are increasing possibilities to propose other ways of producing and supplying food. The example of *Hortec* shows that with a few leaders engaged in the transition and a favoura-ble cultural context, embedded in Catalonia in a long tradition of *pagus* and an interest for good and healthy food, initiatives can lead to a renewed and more sustainable agriculture. This example underlines once more that there is still a place and an interest in agriculture at the edge of the metropoles. And, in another sense, it offers us an interesting perspective, facing the future, on the development possibilities that may exist, beyond the first appearances, in the urban peripheries of all the cities of Europe and the world. That is to say, in those areas with especially complex human and social problems. Areas in which the younger generations could perhaps find, in the context of that renewed agricul-ture, a source of ideas, initiatives and creativity.

In today's globalized world, subject as it is to constant change, it is far from easy to speak with any degree of certainty of what might or might not constitute "strategic models". Indeed, traditionally, agricultural activity throughout the world has been dominated by a significant degree of uncertainty. This is espe-cially true of regions, such as Catalonia and most of the European countries bor-dering the Mediterranean, that have historically constructed agricultural systems on the foundation of small family farms characterized by their highly individu-alistic nature. For this reason alone, the example provided by *Hortec* is clearly both instructive and inspiring, since it highlights values that are fundamental as we seek to defend the future of farming in terms of sustainability, that is, the capacity to adapt, a sensitivity to change and innovation and the promotion of working as part of a broader network that, until now, have enjoyed little accept-ance among the members of the sector.

Clearly, there are no easy answers to the main questions we raise, and which, doubtless, form part of the universe of complexities and uncertainties that we have addressed in the preceding discussion. We should, however, bear in mind the prophetic words of US essayist Scott Sanders (1994) in relation to this uni-verse, when he spoke of the pressing need to link local knowledge with global knowledge. In the meantime, we are sure that the *pagesos* of *Hortec* – who suc-ceeded in turning an idea into a tangible reality – are being guided in their work by this very need.

Acknowledgements

Sincere thanks are due to Iain Kenneth Robinson for his linguistic assistance in the text, and to Albert Santasusagna for his technical support on the map.

This chapter has been prepared as part of the Research Project CSO2015-65787-C6-4-P, supported by the Ministerio de Economía y Competitividad, Government of Spain (MINECO/FEDER, UE), and within the research group GRAM (Grup de Recerca Ambiental Mediterrània), supported by the Generalitat de Catalunya (2017SGR1344).

References

Berger, J. 1991. *Into Their Labours* [Trilogy: *Pig Earth*, 1979; *Once in Europe*, 1987; *Lilac and Flag*, 1990]. New York: Pantheon Books.

Brown, C. and S. Miller. 2008. The Impacts of Local Markets: A Review of Research on Farmers Markets and Community Supported Agriculture (CSA). *American Journal of Agricultural Economics* 90 (5): 1296–1302.

Coromines, J. 1980. *Diccionari etimològic i complementari de la llengua catalana*. Barcelona: Curial Edicions.

Deffontaines, P. 1975. *La Meditérranée catalane*. Paris: Presses Universitaires de France.

Hernández, M., E. Moltó and A. Morote. 2018. Las redes agroalimentarias en la Montaña de Alicante, entre la tradición y las expectativas asociadas a la multifuncionalidad de los paisajes. In *Infinite Rural Systems in a Finite Planet: Bridging Gaps towards Sustainability*, ed. V. Paül, R. C. Lois, J. M. Trillo and F. Haslam McKenzie, 43–50. Santiago de Compostela: Universidade de Santiago de Compostela.

Mak, G. 2000. *Jorwerd: The Death of the Village in Late Twentieth-Century Europe*. London: The Harvill Press.

Morgan, K. 2010. Local and Green, Global and Fair: The Ethical Foodscape and the Politics of Care. *Environment and Planning A* 42: 1852–1867.

Murphy, A. B. 2018. *Geography. Why It Matters*. Cambridge: Polity Press.

Paül, V. and Haslam McKenzie, F. 2010. Agricultural Areas under Metropolitan Threats: Lessons for Perth from Barcelona. In *Demographic Change in Australia's Rural Landscapes. Implications for Society and the Environment*, ed. G. W. Luck, D. Race and R. Black, 125–152. London: Springer.

Pla, J. 1952. *Els pagesos*, First Edition. Barcelona: Selecta.

Pla, J. 1975. *Els pagesos*. (*Obra Completa*, volume VIII). Barcelona: Destino.

Pueyo-Campos, A. 2020. La inteligencia geográfica: construyendo conocimiento 'fidigital' para la Sociedad y sus espacios. In *Desafíos y oportunidades de un mundo en transición desde la Geografía*, ed. J. Farinós, J. Escribano, M. P. Peñarrubia, J. Serrano and S. Asins, 777–792. València: Universitat de València.

Sanders, S. R. 1994. *Staying Put: Making a Home in a Restless World*. Boston, MA: Beacon Press.

Soria, A. 1989. El territorio como artificio. In *Obra Pública* 11: 30–39.

Tort, J. 1998. ¿Posproductivismo en la era de la posmodernidad? Unas reflexiones críticas sobre la realidad actual del medio rural. In *IX Coloquio de Geografía Rural. Comunicaciones*, ed. J. M. Llorente and M. J. Ainz, 191–196. Bilbao: Universidad del País Vasco.

Tort, J. 2006. Del 'pagus' al paisaje: cinco apuntes y una reflexión. In *El paisaje y la gestión del territorio. Criterios paisajísticos en la ordenación del territorio y el urbanismo*, ed. R. Mata and A. Tarroja, 699–712. Barcelona: Diputació de Barcelona.

Tort, J. and R. Català. 2018. Josep Pla's Rural World: A 'Philosophy of Life'? In *Rural Writing. Geographical Imaginary and Expression of New Regionality*, ed. M. Fournier, 45–59. Newcastle upon Tyne: Cambridge Scholars Publishing.

Tort, J. and V. Paül. 2008. From Economic Marginality to Cultural Marginality: The Crisis Afflicting the Farmer's World as Reflected in the Writings of Josep Pla. In *The Global Challenge and Marginalization*, ed. M. M. Valença, E. Nel and W. Leimgruber, 117–132. New York: Nova Science Publishers.

Vicens-Vives, J. 1962. *Notícia de Catalunya*. Barcelona: Destino.

Vilà, J. 1973. *El món rural a Catalunya*. Barcelona: Curial.

Vilar, P. 1964. El medi natural. In *Catalunya dins l'Espanya moderna*, vol. 1, 163–434. Barcelona: Edicions 62.

Woods, M. 2005. *Rural Geography. Processes, Responses and Experiences in Rural Restructuring*. London: Sage.

Yacamán, C., A. Matarán, R. Mata, J. M. López and R. Fuentes-Guerra. 2019. The Potential Role of Short Food Supply Chains in Strengthening Periurban Agriculture in Spain: The Cases of Madrid and Barcelona. *Sustainability* 11 (7): 2080.

11 Rural innovation and the valorization of local resources in the High Atlas of Marrakesh

Fatima Gebrati

Introduction

The High Atlas of Marrakesh is a region with significant potential in natural, human, cultural and economic terms. Although it has long been a lagging locality, it possesses considerable natural and cultural resources which are now being mobilized and valorized in several ways, including the development of rural tourism. In this context, various forms of heritage have become essential components in the establishment of innovative projects which are seeking to develop the region through the promotion of its distinctive identity. This chapter interrogates the relationships between local social and economic vitality and national and global trends. How does the valorization of local distinctiveness affect the dynamics of relations between localities, the region and the State? What are the impacts of these valorization processes? To what extent do authenticity and distinctiveness become conduits for the local creation of wealth and innovation? This chapter analyzes the transformation of these marginal spaces by focusing on the valorization of local identity and rural heritage. The mountain areas of the High Atlas are undergoing a period of profound change; they are devising and experimenting with new models of governance and the management of territorial resources. In this process, local economic systems, local identities and local systems of governance are all being redefined by a series of endogenous and exogenous influences.

In Morocco, mountains have been viewed, traditionally, as marginal and marginalized areas. Therefore, they were excluded from the great productivist transformations of the recent past; the development efforts undertaken by the State were directed toward the more fertile Atlantic plains, to those areas most suited to hydro-agricultural developments and to fuel export (Boujrouf 1996, 37–50; Gebrati 2004b, 6–30). The mountain areas therefore suffer from many handicaps and possess enormous needs, especially in terms of infrastructure and other forms of socio-economic assistance.

This chapter interrogates the evolution, innovation and reclassification of these marginalized areas. In recent years, these processes have succeeded in producing new interactions between public authorities and exogenous actors. This internal and external reconnection of the margins has enabled the emergence of new local products and heritage resources and new socio-spatial identities.

DOI: 10.4324/9781003110095-14

The issue of local development of these mountain areas went unrecognized in Morocco until the State devolved or delegated an increasing share of its responsibilities and powers to various local, interstate or international "authorities" including national and global non-governmental organizations, various local associations, national NGOs and various inter-state organizations (Gebrati 2004b, 40–60). This change resulted in the appearance of new local and external groups of actors as well as new structures.[1] These national changes are inseparable from growth in the scale and influence of international economic and governance systems. While these changes in approaches to development present new challenges, they have also seen the adoption of new strategies which endeavor to combine popular participation and a strengthening of the power of the most vulnerable groups in society through the decentralization of decision-making by public authorities, associations and NGOs. These shifts are producing forms of deliberation, reorganization and mediation which manifest themselves in various ways including multi-organizational networks and systems with "polyarchic" aims and methodologies.

In the High Atlas of Marrakesh, these dynamics have been triggered by a variety of actors from the economic, political and heritage spheres. The valorization of local and heritage resources in Morocco has, to a large extent, been driven by the forces of globalization. As they respond to these forces, the rural and mountain areas of Morocco are experiencing a period of profound change during which they are inventing and experimenting with new models of sustainable development which have the potential to lead to better governance of their territories and better management of their resources.

This chapter aims to analyze the challenges resulting from the transformation of these spaces, with particular reference to the issues of local distinctiveness and rural heritage. This commences with an interrogation of the notion of local distinctiveness in order to understand how the value of these local and heritage resources might be realized through the development of "local products" and rural tourism. This will also involve a consideration of how the valorization of local distinctiveness modifies the dynamics of the relationships between the High Atlas and its hinterlands in both economic and governance terms.

This work is based on a multiple retrospective and prospective approach. It lies at the intersection of three main research themes in geography: the question of development; the territorial mobilization of actors; and the mobilization of territorial resources and innovation in marginalized rural areas. In other words, it seeks to identify to what extent the emergence and the development of tourist activities in rural and mountain areas, through the valorization of local resources, is capable of achieving two ends: the sustainable and equitable revitalization of local economic and social systems; and the preservation and protection of their identities, authenticities, ways of life, histories and heritages.

The methodology adopted here is that of an interpretive case study. Data collection comprised semi-structured interviews with local actors, direct observation of local innovation activities and documentary analysis. Operationalizing the research objectives of this chapter required a balance between quantitative and qualitative inputs in order to understand the various components of

evolutionary, long-term change in the study area. Interviews focusing on innovation and rural tourism were conducted with stakeholders and members of associations and cooperatives in the three valleys, Rhérhaya, Ourika and Zat, in
the High Atlas of Marrakesh. Seventy semi-structured interviews were framed
around the theme of innovation and how the management and development
of local resources could raise local living standards and increase the visibility of
the region.

A brief literature review on the opportunities for rural tourism in remote areas
is followed by a description of the study area which presents a retrospective analysis of mountain tourism as a vector of social and territorial innovation in the
High Atlas of Marrakesh. Subsequently, the processes of harnessing local and
heritage resources are explored, together with a consideration of the changing
relationships between remote rural and urban areas. The conclusion depicts
the fundamental transformations that take place in formerly isolated regions as
they become increasingly connected to the globalized world and emphasizes the
urgent need for governance systems to take the needs of both endogenous and
exogenous actors into account when managing local resources.

Rural tourism: A boon for remote areas?

In the context of the revival of interest in the local and the emergence of new
rural paradigms, rural tourism is a possible strategy for the diversification of rural
economies. New tourism practices can offer an alternative to mass tourism. The
cooperative development of tourism and new perceptions of the countryside can
provide remote rural areas with a basis for development. Nonetheless, if the local
community seeks to benefit from tourism development, local people need to be
major actors in this process (Schmitz 2013). In Morocco, several studies have
traced the development of new rural tourism practices (Boujrouf 1996, 37–50;
Gebrati 2004b, 86–200). Since 2000, tourism in Morocco has experienced an
inflexion; it has undergone profound changes, which have affected not only the
nature of its offerings and the marketing of new destinations, but also the areas
of tourism practices, its organization and the involvement of new players. Berriane and Bernard describe how, up to the 1990s, the sector operated according
to a Fordist model characterized by mass production and product standardization
(Berriane and Bernard 2014, 2–10). Today, post-Fordist tourism is characterized by the non-standardization and flexibility of its products and the promotion
of alternative modes of tourism development. The reality of both demand and
supply shows that more remote tourist destinations are increasing in popularity
and that the populations of these hinterlands are developing new relationships
within their local territories (Berriane and Bernard 2014, 2–10).

The United Nations World Tourism Organization (2008) definition of tourism, as "a social, cultural and economic phenomenon related to the movement
of people to places outside their usual place of residence, pleasure being the usual
motivation", highlights that tourism is a multifaceted phenomenon. This sector
of activity has an economic dimension but also social and cultural ones. Rural
tourism is a form of tourism that can take several forms, dependent upon the

activities and needs being supplied. It allows two worlds to meet, thereby changing the perceptions of city dwellers vis-à-vis rural people and vice versa (Noël
and Willaert 2007). Rural tourism is a sector of activity rooted in locales and
is inseparable from them. There is, therefore, a strong dialectic between rural
tourism and place: on the one hand, the specific places provide rural tourism
with the resources that enable this industry to develop; on the other hand, rural
tourism contributes to the image and identity of these rural places, highlighting their advantages and making them appear more attractive as destinations.
Rural tourism encompasses several types of tourism including cultural tourism
and ecotourism.

Most, if not all, rural locales possess economic, social, cultural and environmental resources which can be mobilized by local actors for development (Pecqueur 2001). These include the history and culture of places, local customs and
crafts, and natural and cultural landscapes.

Local agro-food products and local culinary techniques contribute to distinctive food cultures. These can be harnessed to develop "gastro tourism" attractions
in a world where food consumption patterns are becoming increasingly homog-
enized. Capitalizing on the unique nature of local products is a strategy often
used, individually and collectively, by tourism actors in mountain areas since this
can provide them with a competitive advantage. This strategy of differentiation,
which is based on the competitive exploitation of a local anchorage, has proved
to be popular with local actors and territorial authorities, such as community and
regional associations and national and international NGOs.

The High Atlas of Marrakesh

The study area encompasses four valleys in the High Atlas of Marrakesh; the
Ourika Valley, the Rhérhaya Valley, the Azzaden Valley and the Zat Valley
(Figure 11.1). The inhabitants of these valleys are Berber (Tachelhit)-speaking
Masmodians and share a single cultural identity. This area is both the longest
settled and the most densely populated subregion of the Atlas.

The territorial arrangements of the tribes of the High Atlas are mainly constrained by topography and also by the drainage networks; the territories of most
of the tribes coincide, in whole or in part, with one or more drainage basins
(Bellaoui 1996). The intertwined and inseparable activities of agriculture and
animal husbandry provided the traditional economic base for these mountain
communities.

The High Atlas has some of the highest rates of poverty and vulnerability in
Morocco, and the region lags behind the rest of the country in terms of economic
and social development. One of the main obstacles to economic growth is the
lack of infrastructure, especially in terms of accessibility to upstream villages in
the mountain valleys. Large parts of these areas have therefore remained isolated
and under-resourced which has made both meeting the needs of the population
and promoting development problematic. The domination of the local economy
by informal seasonal activities also inhibits local economic development and the
improvement of local living standards. Nevertheless, the area has experienced a

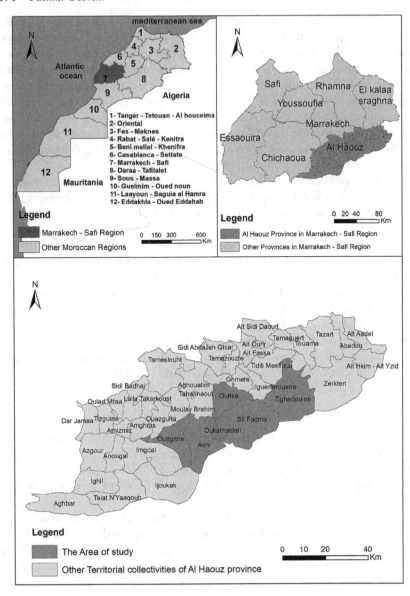

Figure 11.1 The Study Area.
Source: Author.

significant rural transformation in recent decades based on the development of tourism and of orchard production for the national market. In the High Atlas of Marrakesh, a version of sustainable rural tourism is starting to gain momentum, emphasizing local artisanal products, regional culinary varieties and well-rooted seasonal folkloric traditions. However, these changes will only ensure a promising future for this rural region if a tourism-oriented State program of training and human resource development is implemented. Nevertheless, the current growth in tourism demand makes it possible to envisage this mountain region as a "resource" and rural tourism as a tool for the development of the region's resources.

Mountain tourism: a vector of social and territorial innovation

In Morocco, the High Atlas Mountains of Marrakesh are part of the country's marginalized areas. Various policies and strategies have been put in place to improve the living conditions of the local population while also preserving the environment. Multiple territorial initiatives have been focused on the development of respectful, committed and partnership tourism enterprises which have sought to identify, develop and manage new heritage tourism resources. A process of heritagization of local resources is therefore being operationalized in these mountain areas. Both the government and the local people have expectations for tourism. These include the diversification of the economy, the development of new skills, the creation of jobs and the development of a network of stakeholders among which are cooperatives, associations, tourist agencies, state institutions and guides.

It is intended that the global model of income-generating tourist activities being created through centralized and trickle-down processes will be substituted by a decentralized and "sharing" model. The question that arises from such a process is whether this form of tourism development can actually contribute to the sustainable social and economic development of this marginal region.

Mountain tourism began to develop in the 1920s under the colonial government. It benefited from sporting patronage of the mountains, in particular by members of the Casablanca section of the French Alpine Club and by the Marrakesh Tourist Office. The latter played a key role in the establishment of basic infrastructure (lodges and refuges) in the Toubkal Massif, which rises to 4,167 meters. Several sources have confirmed and emphasized the decisive role of road construction in the opening up of the region and the development of tourism (De Mazieres 1937). The creation of the Toubkal National Park in 1942 and the opening of trans-Atlasic traffic routes, in particular the roads connecting Marrakesh (by the Tizi n'Tichka and the Tizi n'Test) and the other urban centers of the southern slope of the Atlas, namely Agadir, Taroudant and Ouarzazate, was critical in providing access to the mountains and therefore to their frequentation by tourists.

Following Moroccan independence in 1956, mountain tourism was promoted by French and Swiss guides, who gave the region a pioneering and outdoor reputation though the very small Moroccan elite tended to prefer the Middle Atlas and the Oukaimeden ski resort rather than the High Atlas. This has resulted in more international tourism in the Rhérhaya Valley and more local tourism in the Ourika Valley. It was only in the 1980s that the Moroccan government began to take real measures to develop tourism in the mountains. Its interventions in this regard were carried out as part of an intersectoral pilot experiment in rural high mountain economies, which was a Moroccan–French cooperative venture undertaken in the Central High Atlas (Moudoud 2000, 35–36).

Since 2002, Morocco has embarked on a policy of enhancing its cultural heritage to improve the visibility and to develop the touristic attractiveness of its regions. This has resulted in the creation of a model for developing tourism activity through the integration of heritage processes and the acknowledgment of local identities. Morocco opted for the approach "Pays d'Accueil Touristique" (PAT) in its Vision 2010 and Vision 2020 as well as in its Rural Tourism Development Strategy (Ministère du Tourisme 2010). These projects aim to equip localities with coherent and legible tourist identities (PATs) and to develop "tourist niches", such as the practice of agritourism, which are linked to specific natural or cultural resources and are intended to appeal to specific clienteles.

However, despite these efforts, mountain tourism remains a "niche product" (Moudoud and Ezaïdi 2006), which is not yet comparable with the demands for seaside and urban tourism in Morocco. Consequently, tourist activity in the mountains is dependent on the cities (Casablanca, Marrakesh, Agadir, etc.), which generate international tourism flows and are intermediate destinations and national ports of entry for international tourists who wish to visit the mountain regions.

Nevertheless, the tourism sector does contribute to the achievement of local development in the High Atlas (Gebrati 2004a), since the majority of the population benefits either directly or indirectly from the revenues of the sector which provides an extra source of regional income. However, a question remains as to whether the growth of rural tourism can decisively revitalize an area whose economy has always been dominated by traditional agricultural and livestock production and to what extent it can serve as a primary basis for its development following the example of mountain tourism in developed countries such as Switzerland and Austria.

Although, an increase has occurred, in terms of tourist infrastructure, with the construction of mountain huts, and assistance has been provided to develop homestay accommodation, these initiatives have failed to achieve all the hoped-for local economic development objectives. Although tourism has provided some local financial benefits, on average less than 15% of the amount spent by foreign tourists coming to climb or trek in the region remains in the area and this accrues mainly to the guides who constitute a small minority of local population.

Indeed, the sale and organization of tours and accommodation are mainly conducted outside the High Atlas, either in France or in travel agencies in the major Moroccan cities. Local tourism therefore benefits firstly foreign or non-local travel agencies and tour operators, and only secondly guides, mule drivers and owners of lodgings who are drawn from the local elites. Thus, the circuits of tourist expenditure remain geographically limited. Some communities and individuals in the Ourika and Rhérhaya Valleys have been able to take advantage of their proximity to Marrakesh. However, several valleys in the mountainous hinterland of Marrakesh have not been promoted as tourist destinations and have failed to take advantage of their potential. Indeed, ever since the establishment of the French Protectorate of Morocco in 1912, official attention has been focused on the valleys surrounding Toubkal National Park. Furthermore, most lodges are located in the valleys from which mountain guides were first recruited.

Even in those valleys with strong tourist patronage, conflicts exist between the villages housing numerous guides and those which possess neither guides nor tourist accommodation. The benefits of tourism are therefore experienced in a limited set of localities, particularly in villages close to roads or to well-established trekking routes.

Albeit at varying rates in its different localities, the economy of the High Atlas is currently shifting from a near complete dependence on agricultural, pastoral and forestry activities as several local actors realize the tourism potential of the region's natural and cultural heritage. In this socio-spatial repositioning, tourism becomes a complementary activity which revalorizes agriculture as heritage. This leads to the creation of new tourism products, a process which can be destabilizing and can call into question the authenticity of local craft products and the tourist experiences that are being conveyed thereby. Boujrouf (2014) observes that it is quite obvious that these specific and heritage resources are starting to impose themselves in certain Moroccan peripheries and in particular the High Atlas and the Great South as alternative resources which value local knowledge and reconstruct local identities based on heritage.

These economic and employment shifts are contributing to the acceleration of the transformations of local mountain societies and of their production and representation systems. The potential of tourism as an exit strategy from rural marginality and poverty is fully acknowledged by the public authorities as they seek to diversify an economy overly dependent on agriculture and identify new forms of development such as rural and heritage tourism (Boujrouf 2014).

The activation of territorial resources

In the High Atlas of Marrakesh, the use of a range of resources, including local heritages, has enabled different actors to mobilize around new projects and initiatives. Heritage is being used to showcase local distinctiveness through both agricultural and tourism-related development projects. One of the main changes in the landscape of agricultural production is that arboriculture is becoming increasingly dominant; terraces are constantly being constructed to increase the crop growing area in order to meet the needs of a growing population. The new terraces are mainly for the intensive cultivation of fruit trees. Species such as apple, cherry, peach and quince are preferred because of their rapid growth rates. These new terraces are mainly near the villages, to facilitate regular watering and close monitoring of crop growth. A recent initiative by a leading local landowner, in the lower Rhérhaya Valley, has been the introduction of kiwi fruit production in greenhouses and using drip irrigation. This innovation has now been replicated in other valleys of the Atlas. Over the last 30 years, the introduction of rosaceous (cherry, apricot, apple, pear, etc.) arboriculture has been a real revolution in the agricultural system, especially in the valleys on the northern slope of the High Atlas, such as the Rhérhaya, Azzaden and Ourika, where it has changed the local economy from subsistence to a market basis.

The social and territorial innovation observed in recent years has been gaining momentum. The annual Asni Township Nut Festival provides one example of the relationships between the various actors and specific mountain localities.[2] This festival is the result of a partnership between the association "Friends of the Toubkal National Park for the protection of the environment and walnut", the province, the town of Asni and the cooperative Idraren nut producers. This festival is a springboard used by the local population to highlight its varied potential and to market its local products. According to the High Atlas Foundation[3] (2019), "over 6 tons of organic walnuts have been processed and 600 liters of organic walnut oil have been produced to date". The processing unit, created by the Ministry of Agriculture and the High Atlas Foundation and co-managed by the Idram cooperative, is the first factory of its type in Morocco. Its products are sold on the local market and internationally. Its profits allow this municipality to support education and water projects, and initiatives to benefit women and young people.

These local products and initiatives contribute to the diversification of the local economy thereby creating wealth, reducing poverty and protecting local areas from the instabilities of globalization and the potentially impoverishing process of standardization. Furthermore, the territorial heritage resources are becoming increasingly attractive to tourists who thereby become participants in the resource activation process. This happens through the identification of these localities as tourist destinations and then through the promotion and consumption of local products, both directly by the tourists and indirectly through export markets.

Local products can be differentiated on the basis of their point of origin and guaranteed through a geographical indication (GI) which provides them with quality assurance and a form of authenticity. By anchoring these products to their local place of origin it becomes possible to establish a virtuous circle of quality linked to this territory. Indeed, the foregrounding of this local specificity can generate positive outcomes in economic, social and environmental terms. It can help to create or consolidate supply chains and to provide local products with an appearance of exclusivity which is desired by many urban consumers. As shown in Figure 11.2, the Economic Interest Grouping (EIG) project in the Oued Zat

LOCALISATION DE L'UNITÉ	COMMUNE RURALE AÏT FASKA : PROVINCE AL HAOUZ
COMMUNES RURALES CONCERNÉES	AITFASKA, TIDILI MESFOUIFA, TAMAGUERT, AIT HKIM ET TIGHDOUINE
TYPE D'ACTIVITÉ	TRITURATION ET MISE EN BOUTEILLE DE L'HUILE D'OLIVE EXTA VIERGE
CAPACITÉ DE TRITURATION	60 TONNE/JOUR
HUILE D'OLIVE PRODUITE	600 TONNES/AN
BENEFICIAIRE DU PROJET	GIE BASSIN OUED ZAT
COOPERATIVES MEMBRES	8 COOPERATIVES AGRICOLES
NOMBRE BENEFICIAIRES	3 200 AGRICULTEURS
SUPERFICIE OLEICOLE	4 400 HA
PRODUCTION	5 000 Tonnes
Coût du projet (construction, equipement, matériel hors chaine de trituration)	17.100.000 MAD
DATE D'ACHEVEMENT DU PROJET	SEPTEMBRE 2013

Figure 11.2 Economic Interest Grouping (EIG) in the L'Oued Zat Basin.
Source: Author.

Basin has promoted local olive oil and succeeded in penetrating national and international markets. The successful promotion of local products also has the potential to stimulate integrative development projects at the local level which can assist in sustaining local cultures and environments.

In this context, the development of local products such as argan oil and saffron (Boujrouf 2014) and community practices such as the agdals, a traditional land management practice of communal rangelands (Auclair and Alifriqui 2013), contribute to the creation of new tourist circuits preserving local heritages and strengthening new economic sectors.

Territorial innovation and new relationships between rural and urban areas

The proximity of the metropolitan cities of Casablanca and Marrakesh, which constitute the major markets for the sale of these products, has encouraged set-tlers, emigrants and city dwellers including leading citizens, to invest in the introduction of rosacea crops. The city of Marrakesh has been a catalyst for this change at regional and national levels (Boujrouf 2005).

The influence of the city on the economy of the valleys as well as the degree of evolution and modernization of the systems of production and culture has not been uniform across all the valleys of the study area. This variation results from both the ruggedness of the terrain and the nature of transport facilities available.

The High Atlas of Marrakesh has now become a space of exchange that feeds on large modern cities such as Marrakesh, Agadir and Casablanca. These exchanges occur in the cultural sphere as well as in the movement of goods. These exchanges have bought about the expansion of agricultural areas at the expense of forest and an improvement in rural living standards. The city is therefore a vector of change in the marginal spaces of Morocco such as the High Atlas as a result of its participation in the production and the emergence of a modern and capitalist economy. This section considers how these two areas coexist.

The increasing economic complementarities between Morocco's major cities and its rural regions are bringing about economic and social disruptions on the country's margins. As a result of this shift, the flow of people and goods between urban and rural areas is becoming increasingly important with marginal rural spaces experiencing both outflows and inflows of population. These migration flows are occurring both between the upstream and downstream portions of the High Atlas of Marrakesh and between the city of Marrakesh itself and the valleys of the High Atlas. The new city/countryside relationships established through these movements are producing new rural/urban connections (Barcus and Brunn 2010); including income transfers, and other interlinked economic activities. As the rural spaces in the study area change as a result of increased mobility multiple and diverse land uses and modern and traditional practices still coexist.[4]

At this stage of rural transformation, the growing movement of tourists into the study area is a vital component of the change process. Tourist interactions with the local population not only have an economic effect but also have a social impact, particularly on conceptions of local identity. The representations of local cultures and landscapes brought about, at least in part, by the interactions between the local population and tourists stimulate new forms of cultural expression highlighting local identities, such as rural museums, but it also disturbs traditional local values, such as those relating to the role of women.

Discussion and conclusion

Today the High Atlas of Marrakesh[5] is becoming integrated into trade circuits and is opening up to the rest of the country. This is altering the "center/periphery" relationships and introducing secondary hierarchies within the study area. The inclusion of the High Atlas of Marrakesh in the tourist circuits is intensifying relations not only with Morocco's metropolitan areas, and especially with the city of Marrakesh, but also with Europe. This multiplication of relationships, not only between the various actors but also between the various local and global scales, is creating new forms of activity and contributing to the emergence of new linkages.

Social, economic, cultural and even political implications can be discerned behind this local process of touristification. While tourism often appears to be a real engine of development, it possesses diverse and sometimes contradictory logic since it creates economic wealth while simultaneously altering the traditional cultures and landscapes that it purports to valorize.

Therefore, even if the advent of tourism has made the reinvention of the High-Atlas space possible, its "touristification" on the economic, as well as social planes, poses problems. While some sectors such as arboriculture and beekeeping have been affected by the development of tourism, for most traditional activities this does not seem to have been the case and the impact of tourism on traditional crafts and agro-pastoralism remains limited. Handicrafts and agriculture, and even trade to a large extent, remain largely outside the new economic circuits generated by tourism: the supply of foodstuffs and the production of domestic equipment and even souvenirs (carpets and pottery for example) continue to take place in the major towns.

Tourism results from openness. It can enhance local cultures, improve the living conditions of local populations and open up rural areas. But it has often led to changes in lifestyles and to new practices which sometimes conflict with traditional values as rural populations experience monetary enrichment and adopt more urban lifestyles.

In the High Atlas, as elsewhere, tourism has resulted in the creation, of two totally different "spaces" which, at least potentially, are not complementary but conflictual. The resulting need for coexistence between two different social and even cultural groups occupying the same small territory can constitute an additional source of frustration. Moreover, tourism is an activity largely dependent on the outside world for its existence and it remains, at least in the High Atlas, a speculative sector controlled by urban investors and promoters.

More broadly, rural innovation is a process whereby a multiplicity of actors, with often divergent, if not contradictory, aims, are mobilized to engage in localized social dynamics, in a context of globalization. Rural innovation in the High Atlas of Marrakesh is highly dependent on both endogenous and exogenous actors; the introduction of modern and novel economic activities has changed both the landscape and a local economy which was formerly dominated by the traditional activities of agro-pastoralism and crafts. Currently, the space of the High Atlas of Marrakesh is undergoing demographic, economic and social development. This evolution encompasses an increasingly important upstream–downstream relationship with the cities of the nearby plains (Marrakesh and Casablanca) and the intermountain areas. These comings and goings, with the social exchanges which they induce, have brought about changes in behavior and spatial practices. Agricultural practices are being modernized, and the modes of trade are becoming more varied. This leads to different forms of occupation of the land and increases the nature and volume of the exchange of goods with the cities. It also opens up communication between the Amazigh people and the urban world.

Since the beginning of this century, the mountain areas that were formerly bypassed by the great productivist transformations are no longer perceived as backward and immobile environments but, on the contrary, they have become laboratories of territorial innovation. They are engaged in a period of profound change; they are experimenting with new models of governance of their territorial resources and of sustainable development. The Atlas territory and its actors are not immune to the changes that permeate global society. They are affected by modifications of both their socio-economic structure and its dynamics. Indeed, the globalization process has brought about new individual and collective socio-spatial configurations. These new configurations require the readmission and redefinition of the actors within the territory. Local and external actors seeking to carry out projects will need to call on people with varied skills and resources. These will include community groups, elected officials, professionals and development agents. To reaffirm their individual and collective identity, these actors will claim legitimacy based on their connections to a particular social group or space.

The High Atlas of Marrakesh is today thought of and organized as a resource space. The deep transformations of the economy and its relationships with mountain environments (exogenous factors) and the mutations of the mountain societies themselves (endogenous factors) are leading to the emergence of new forms of social and spatial organization of these spaces.

Within this space, concepts of culture and heritage are being employed to forge a social identity specific to the territory and thus to promote its distinctiveness. With the advent of globalization and of related phenomena such as the increase in the number and range of actors, many with diverse aims and ideologies, carrying out local development projects through local social and territorial systems has become more complex. The local/global dichotomy, on which this chapter is based, is not a simple opposition; they are two sides of the same coin. These two processes, globalization and a return to the local, can be seen as complementary. This renders the issues of territorial restructuring and the territorialization of actors, activities and public action as comprehensible if more complex. Nevertheless, it is through the lens of complementarity between local and global actors that the articulation of public policies, regional planning and territorial development needs to be considered.

Notes

1 Such as associations, cooperatives and NGOs.
2 Its slogan is "The natural local product at the service of mountain tourism". The objective of this festival is the preservation and enhancement of the walnut tree as a cultural and heritage symbol and one of the essential heritage components of the arboreal area of the province of Al-Haouz.
3 https://www.highatlasfoundation.org/media-haf/haf-in-the-news
4 In this sense, Barcus and Brunn (2010) introduce the concept of place elasticity as a new conceptualization of place attachment made possible by innovations in communication and transportation technologies. Place elasticity allows individuals to live in distant locales while maintaining close interaction with a particular place.
5 Administratively speaking, the High Atlas of Marrakesh is part of the province of Alhaouz, which includes 136 classified accommodation establishments, comprising 2,362 rooms with an estimated capacity of 5,421 beds, in addition to a significant number of hostels.

References

Auclair, L. and M. Alifriqui. 2013. *Agdal, Patrimoine Socio-écologique de l'Atlas Marocain.* Paris: IRD.

Barcus, H. R. and S. D. Brunn. 2010. Place Elasticity: Exploring a New Conceptualization of Mobility and Place Attachment in Rural America. *Geografiska Annaler, Series B. Human Geography* 92: 281–295.

Bellaoui, A. 1996. Tourisme et Développement Local dans le Haut-Atlas Marocain: Questionnement et Réponses/Tourism and Local Development in the Moroccan High Atlas Mountains: Some Questions and Answers. *Revue de Géographie Alpine* 84 (4): 15–23.

Berriane, M. and M. Bernard. 2014. *Le Tourisme dans les Arrières Pays Méditerranéens. Des Dynamiques Territoriales Locales en Marge des Politiques Publiques.* Agdal: Ed. Université Mohammed V.

Boujrouf, S. 1996. La montagne dans la politique d'aménagement du territoire du Maroc/ The place of mountain areas in Morocco's national planning and development policies. *Revue de géographie alpine,* 84(4): 37–50.

Boujrouf, S. 2005. Innovation et Recomposition Territoriale au Maroc: une Mise en Perspective Géo-historique. In *Le Territoire est Mort: Vive les Territoires!: Une Refabrication au Nom du Développement,* ed. B. Antheaume and F. Giraut, 133–156. Paris: IRD (37–50).

Boujrouf, S. 2014. Heritage Resources and the Development of Tourist Areas in the High Atlas and Southern Regions of Morocco. *Revue de Géographie Alpine* 102 (1): 1–14.

De Mazieres, M. 1937. *Promenades à Marrakech.* Paris: Horizons de France.

Gebrati, F. 2004a. Le Tourisme et les Conceptions de la Durabilité à Travers les Actions du Développement Local dans le Haut-Atlas Occidental. In *Le Tourisme Durable, Réalités et Perspectives Marocaines et Internationales*, ed. R. Bousta, F. Albertini and S. Boujrouf, 317–326. Marrakesh: Centre de Recherche sur les Cultures Maghrébines, Faculté des Lettres et Sciences Humaines, Université Cady Ayyad.(317–320).

Gebrati, F. 2004b. La mobilisation territoriale des acteurs du développement local dans le Haut-Atlas de Marrakech. Thèse de géographie. Université Joseph Fourier de Grenoble, (6–30).

Ministère du Tourisme. 2010. Vision stratégique de développement touristique. Contrat programme 2011-2010. Accessed June 10, 2021. https://www.icao.int/meetings/moroccan-economic-forum/documents/tourismmaroc.pdf

Moudoud, B. 2000. Production et gestion du tourisme de montagne au Maroc. Doctoral thesis, Université Joseph Fourier, Grenoble 1.

Moudoud, B. and A. Ezaïdi. 2006. Le Tourisme National au Maroc: Opportunités et Limites de Développement. *Téoros* 25: 25–30.

Noël, G. and E. Willaert, eds. 2007. *Georges Pompidou, une Certaine Idée de la Modernité Agricole et Rurale*. Brussels: Peter Lang.

Pecqueur, B. 2001. Qualité et Développement Territorial: l'Hypothèse du Panier de Biens et de Services Territorialisés. *Economie Rurale* 261: 37–49.

Schmitz, S. 2013. Dutch Vision of Community Based Tourism: Dutch People in the Belgian Ardennes. In *The Sustainability of Rural Systems: Global and Local Challenges and Opportunities*, ed. M. Cawley, A. M. d. S. Bicalho and L. Laurens, 218–225. Galway and Okayama: Whitaker Institute, NUI Galway and CSRS of the International Geographical Union.

United Nations World Tourism Organization. 2008. Understanding Tourism: Basic Glossary. Accessed June 8, 2021. http://media.unwto.org/en/content/understanding-tourism-basic-glossary

12 Does an agricultural products' certification system reorganize vegetable farmers? A case of the VietGAP program in Lam Dong Province, Vietnam

Doo-Chul Kim, Tuyen Thi Duong, Quang Nguyen, and Hung The Nguyen

Introduction

Food safety is an important issue in the governance of food chains, especially in the globalization era where food can take an incredibly lengthy journey from farm to table. While studies on food safety proliferate in developed countries (Hu et al. 2019, 268), attention to this subject is inadequate in the developing world, except in the case of some export of agricultural products (e.g., coffee, shrimp, or fish). However, a thorough transformation of the food system in developing countries has been recently observed, which was driven by new food safety requirements and other demand-side factors (Reardon et al. 2019, 47). This chapter analyzes the reorganization of supply chains toward producing safer agricultural products with a case study of safe vegetable production in Vietnam.

In recent years, there has been an increasing demand for safe food, especially safe vegetables, among consumers in Vietnam due to widespread foodborne illnesses. In Vietnam, safe vegetables is defined as "fresh vegetables produced and preliminarily processed in accordance with food hygiene and safety regulations in VietGAP (Vietnamese Good Agricultural Practices) or other GAP standards equivalent to VietGAP" (Ministry of Agriculture and Rural Development 2008, 2). In the context of Vietnam, the term does not imply organic vegetables but refers to vegetables produced with fewer pesticides and fewer chemicals. It is considered to mean the same as the Vietnamese Good Agricultural Practice (VietGAP) vegetables sold at a modern supermarket. Vietnamese are well known for their dietary habit of consuming a lot of raw vegetables, especially leafy vegetables. However, along with rapid agricultural development since the 1990s, Vietnamese farmers have sharply increased the use of pesticides and chemical fertilizers. Low import taxes on agrochemicals and the increased supply from domestic agrochemical companies have promoted their use (Nguyen et al. 1999, 1). The highest amount of pesticides per hectare is used in the production of vegetables. This, in turn, has resulted in many incidents related to food poisoning since the late 1990s, causing consumers to have great concerns about

DOI: 10.4324/9781003110095-15

pesticide residues. This rising demand for safe vegetables has also been accelerated by growing incomes (Wang, Moustier, and Nguyen 2014, 13) and a rapid
blooming of supermarkets (Reardon et al. 2003, 1144–1146).

Various public and private governance measures regarding food safety have
been established to improve the agro-food industry, which, until the early 2000s,
was dominated by a traditional value chain running from producers to collectors,
then to wholesale markets, and finally to traditional retailers (Cadilhon et al.
2006, 35–37). Public policy has been evolving around the trend of retail modernization or "supermarketization", which relies on the assumption that rational
consumers will shift toward the supermarket channel because of food safety
concerns as well as its suitability with the modern lifestyle (Wertheim-Heck,
Vellema, and Spaargaren 2015, 95–96). This public effort consists of developing
modern food retail facilities (e.g., supermarkets and high-quality stores) as part of
the urbanization process and the establishment of related regulations (e.g., Law
on Food Safety (National Assembly of Vietnam 2010); provincial food safety
guidelines).

Besides these public efforts, private measures to improve quality standards in
the agro-food value chain have also emerged on an international scale as food
chains are gradually globalized under the increasing influence of transnational
agro-enterprises and other organizations. This includes not only a wide range of
international quality standards in food production but also various financial and
technical support for different stakeholders in the value chains, especially smallholders, who face high barriers to market participation (Tennent and Lockie
2013). In Vietnam, the operation of transnational food producers such as the
Charoen Pokphand Group and the application of standards such as Best Aquaculture Practices (BAP) or Aquaculture Stewardship Council (ASC) Standards
in the production of fruits, vegetables, poultry, pork, shrimp, and fish have set
new benchmarks for the whole industry. In fact, the practice of standards such
as GLOBALGAP has been the inspiration for the establishment of the national
GAP in Vietnam and many other Asian countries (Nabeshima et al. 2015,
11–27).

These new policies and market movements toward higher food safety are
expected to induce upstream value chain actors, such as farmers and middlemen
(intermediaries), to transform the way they do their business (Nicetic et al. 2010,
1900). However, previous studies have shown that small farmers encounter many
market entry barriers. In adopting high standards, small farmers in developing
countries often struggle with the cost of compliance and access to capital and
finance (Ouma 2010, 199; Tennent and Lockie 2013, 170–171). In many cases,
besides the high cost of compliance, the benefit of producing vegetables under
high-quality standards may be ambiguous to farmers. Strict quality control often
leads to a situation in which only a small fraction of the outputs can be sold to
high-value retailers at premium prices (Hoang 2018, 147–152). A recent study
in Hanoi, Vietnam found that about 80% of "safe vegetables" are still sold in
traditional markets at a price almost similar to that of uncertified products (Ngo
et al. 2019, 364). Consumers' familiarity with food quality standards was low,

which lowers their willingness to pay for high-quality products (Nguyen et al. 2017, 76–78; 2018, 79). In some cases, small farmers are often reluctant to switch to the new form of production because they want to maintain long-lasting relationships with traditional collectors and other actors in the existing value chain (Hoang 2018, 150).

It is also difficult to make middlemen comply with the new quality standards in the value chain. The middlemen system in less developed agricultural value chains is usually cumbersome and difficult to regulate. Previous studies have shown that the Vietnamese agro-food industry is filled with a large number of middlemen, operating actively in various value chains of products such as vegetables (Cadilhon et al. 2003), rice (Reardon et al. 2014), tea (Tran and Goto 2019), and shrimp (Tran et al. 2013; Van Duijn, Beukers, and Van der Pijl 2012). This crowded, flexible group is necessary for collecting products from a highly fragmented farming sector full of small farmers. However, they make the matter of quality control complicated. On the one hand, middlemen are important in quality control in existing value chains, which makes wholesalers prefer trading with them than directly with farmers (Abebe, Bijman, and Royer 2016, 209–210). On the other hand, when new quality standards come into play, they may hinder the enforcement of quality control by lowering traceability (Ouma 2010, 214–215). While conducting a study on safe vegetable production in the Red River Delta, Pham et al. (2009, 384) noted that the collectors, who were frequently the heads of a safe vegetable cooperative, often mixed the product of farmers in their cooperative with those outside to obtain a higher quantity, thus contributing to noncompliance. Another example can be found in shrimp production in the Mekong Delta, Vietnam. Tran et al. (2013, 6) observed that shrimp farmers sold their outputs in small quantities to middlemen, who had to travel by boats to access these farms, and then the collection process continued through many levels of middlemen, which led to the mixing of low-quality (e.g., contaminated with antibiotics) with high-quality shrimp. The same situation has also been observed in other value chains of products such as citrus fruit (Nicetic et al. 2010, 1896) or rice (Niewöhner et al. 2016, 81).

The above-mentioned difficulties in the production and distribution processes inevitably hamper the spread of high-quality standards in Vietnamese agro-food value chains. However, some ongoing trends are being observed in the way farmers and middlemen operate in the new value chains with higher quality standards. Small farms tend to unite into groups or cooperatives. This has many benefits such as lower costs of certification and compliance. It is also easier to satisfy strict requirements in terms of both quantity and quality, collective quality control, service provision, and mutual support (Moustier et al. 2010, 76–77; Wang et al. 2012, 351). In the long term, changes in farming are accompanied by rural factors of market transformations (e.g., improved markets for rural labor, inputs and extension services, credit, land) and intensification (Reardon and Timmer 2014). In addition, there has been a trend of disintermediation, in which the operation of midstream actors such as middlemen is increasingly sidelined, and wholesalers buy products directly from farmers or cooperatives (Reardon 2015, 57–59; Reardon et al. 2014, 5). Disintermediation leads to higher traceability and, therefore, improves quality control in the value chain. With these trends, a thorough reorganization of the Vietnamese food supply chain is in progress, bringing about substantial innovations in the upstream production sites, the rural areas.

The VietGAP is a major program introduced by the Vietnamese government in 2008 to tackle unsafe vegetable production and its related social issues. It consists of standards and guidelines for the certification of safe fruits and vegetables with 65 criteria separated into 12 main categories: site assessment and selection, planting material, soil and substrates, fertilizers and soil additives, water, chemicals, harvesting and handling produce, waste management and treatment, workers and training, documents, records, traceability and recall, and internal audit and complaint handling.

However, after 10 years of implementation, the achievement of the VietGAP program is still limited. As of 2016, the VietGAP's certified area in Vietnam was 4,355 hectares, which accounted for only 0.48% of the total planting area. However, the case of Lam Dong Province, the center of vegetable production in Vietnam, stands out. The VietGAP certified area for vegetable cultivation in Lam Dong Province was over 1,548 hectares in 2016, which accounted for 36% of the total VietGAP vegetable area nationwide.[1] As the certified areas and the number of farmers of the VietGAP is rapidly increasing in this province, a new distribution channel has been established with a reorganization of local farmers.

Here, we aim to clarify how the agricultural certification system reorganizes vegetable farmers with a case of implementation of the VietGAP program in Lam Dong Province, Vietnam. In particular, we examine the characteristics of the VietGAP vegetable producers and their transformation from conventional farmers. The study site is the Duc Trong district, one of the most concentrated vegetable production areas in Vietnam (Figure 12.1). The field survey was

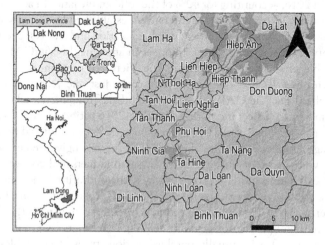

Figure 12.1 Location of the study site.

Source: Created by authors.

conducted in June 2019. The survey, using a semi-structured questionnaire, was conducted with the participation of the 34 VietGAP farmers.[2] The sample was chosen based on discussions with the agricultural officers in the Duc Trong district, which includes all villages carrying out the VietGAP production. In addition, 41 conventional farmers were also randomly approached and interviewed using a similar questionnaire, the results of which can be compared to those from the VietGAP sample. In addition, all the VietGAP buyers[3] in the Duc Trong district, who collect the VietGAP vegetables from farmers and distribute them to supermarkets, were also interviewed.

Outline of the study area

Lam Dong Province is located in the Central Highlands of Vietnam, about 300 km from Ho Chi Minh City and 1,300 km from Hanoi, the capital city (Figure 12.1). Lam Dong Province benefits from numerous favorable conditions for vegetable production, in which the climate and soil are suitable for tropical, subtropical, and temperate vegetables. Agricultural land accounts for approximately 38% of the total area, 63% of which is covered by red soil suitable for vegetables and industrial crops. Lam Dong Province is 800–1,500 m above sea level, which results in year-round stable weather conditions with an average temperature of 18°C–25°C. With the largest vegetable production area in Vietnam, the province is a major supplier of vegetables to urban areas such as Ho Chi Minh City and Hanoi. Moreover, Lam Dong Province is also the leading region in safe vegetable production, with 36% of the total VietGAP certified area for vegetable production in Vietnam.

The Duc Trong district is in the southeast of Lam Dong Province, 26 km from the city of Da Lat, the provincial capital (Figure 12.1). The district is one of the largest vegetable cultivation areas in Lam Dong Province with 19,857 hectares in 2016, accounting for 40% of the total arable land. In addition, many agricultural products purchasing centers were established by major Vietnamese supermarket companies due to the favorable location and a developed transportation network. Duc Trong is also a pioneer in applying the VietGAP standards in vegetable production in Vietnam. The VietGAP program was introduced in Duc Trong in 2009 as one of the pilot cases, just one year after the policy was launched. The VietGAP certified areas in Duc Trong increased nearly nine times from 43 to 382 hectares during the 2012–2016 period. In 2016, 55% of the certified areas were cultivated by farmer groups, 22% by individual farmers, and 23% by agribusiness companies (Figure 12.2). Interestingly, the certified vegetables in Vietnam are mainly produced by small farmers and cooperatives, not by large-scale farms and enterprises. In fact, the authorities encouraged small farmers to establish cooperatives for the VietGAP production by subsidizing the cost of certification. Farmers also need to cooperate to cope with the disadvantages of the small production scale in dealing with buyers (e.g., intermediaries and supermarkets), who usually purchase various types of vegetables in large quantities.

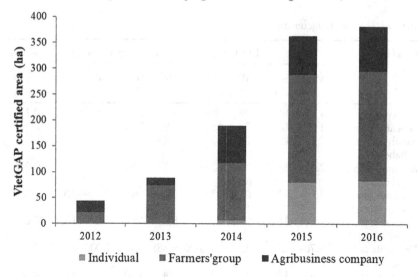

Figure 12.2 Changes in the VietGAP certified area in the Duc Trong district.

Source: Agroforestry and Fisheries Quality Management Department of Lam Dong Province.

Who are the VietGAP farmers?

To clarify the impact of the VietGAP implementation on vegetable farmers in the Duc Trong district, this study compares the ones adopting this standard in their cultivation (the VietGAP group) with those following the conventional way (the conventional group) in terms of their demographic and cognitive characteristics, social networks, and market access.

Basic characteristics of the VietGAP farmers

Table 12.1 shows that the VietGAP farmers, on the one hand, are significantly different from the conventional farmers in terms of education, family laborers, the number of regular employees, and the proportion of income from vegetable cultivation in total household incomes. On the other hand, age, amount of seasonal hired labor, and vegetable cultivation acreage were not significantly different between the two groups. The VietGAP farmers have higher education levels than the conventional ones. Needless to say, the former with higher education tend to adapt easily to the new cultivation standards as well as access new markets. In addition, the proportion of income from vegetable cultivation in total household incomes among the conventional farmers was higher at 91% (compared to 76% in the VietGAP group). The high dependence of household incomes on vegetable cultivation might be a barrier that prevents them from taking risks and shifting from conventional farming practices to safe vegetable production.

Table 12.1 Farm characteristics

Variables	The VietGAP farmer				Conventional farmer				p-Value[1]
	Mean	Std. Dev.	Min.	Max.	Mean	Std. Dev.	Min.	Max.	
Age (years)	45.5	10	29	65	48.6	11.3	27	71	0.22
Education (years)	11.6	2.77	7	17	8.98	2.64	3	16	0.00*
Family labors (laborers)	2.5	1.21	1	7	1.93	0.79	1	5	0.02*
Regular employees (laborers)	2.21	3.98	0	20	0.17	0.59	0	3	0.00*
Seasonal hired labor (days)	38	60.8	0	300	15.9	36.5	0	200	0.06
Vegetable incomes (%)[2]	76	21.2	37.5	100	91.2	15.7	50	100	0.00*
Vegetable acreage (m[2])	9,704	8,293	2,000	39,000	9,585	8,464	2,000	5,0000	0.95
Vegetable plots (plots)	2.94	1.81	1	9	2.46	1.82	1	12	0.26
No. of observations	34				41				

Source: Authors' survey.
Note: [1] p-Value of two-sided t-test on the equality of means; * significantly different at 5%.
      [2] Percentage of income from vegetable cultivation in the household's total income.

Regarding the use of labor, although there were no significant differences between the two groups in terms of the amount of seasonal hired labor, the data indicated that the VietGAP households tended to employ more family laborers and hire more workers on a regular basis. This could be explained by the fact that the VietGAP vegetable production requires more labor to fulfill technical requirements during cultivation. The VietGAP farmers also need additional laborers to sort and pack their products before handing them to buyers (or intermediaries).

The comparison of vegetable cultivation acreage does not show any significant differences between these two groups, suggesting that the scale of farming is not likely to be a determinant of whether a farmer participates in safe vegetable production. However, it should be noted that this region is one of the main vegetable growing areas in Lam Dong Province and in the whole country, and the cultivation area is, in fact, quite large in these groups compared to the average scale of farming in Vietnam.

Contract-based vegetable cultivation

Survey data show that 94% of the VietGAP farmers have a long-term contractual relationship with their buyers, which helps them to guarantee a stable price for their products (Table 12.2). This is considered to be one of the most obvious benefits that farmers receive from converting to the VietGAP farming.

Table 12.2 Contract and price guarantee

	Contract with buyers		Guarantee of stable price	
	Yes	No	Yes	No
The VietGAP farmers	32	2	32	2
(n = 34)	(94%)	(6%)	(94%)	(6%)
Conventional farmers	8	33	7	34
(n = 41)	(20%)	(80%)	(17%)	(83%)

Source: Authors' survey.
Note: The percentages are in parentheses.

Table 12.3 Crop choice decision and harvesting process

	Crop choice		Harvesting and packaging		
	Ordered by buyers	Farmer decides	No. sorting and packaging	Sorting only	Sorting and packaging
The VietGAP farmers	32	2	13	13	8
(n = 34)	(94%)	(6%)	(38%)	(38%)	(24%)
Conventional farmers	6	35	40	1	0
(n = 41)	(15%)	(85%)	(98%)	(2%)	(0%)

Source: Authors' survey.
Note: The percentages are in parentheses.

However, to enjoy these benefits, they are also required to comply with the Viet-GAP standards as well as the buyers' specific requirements such as crop choices, cultivation schedule, and keeping logs of the types and quantities of pesticides and chemical fertilizers used during the cultivation seasons. In our sample, 94% of the VietGAP farmers answered that they chose the types of vegetables according to the orders from buyers (Table 12.3). In addition, to meet the demand from supermarkets, the VietGAP farmers should diversify the vegetables grown on their farms and supply them to buyers on the requested delivery date according to the requirements on quantity and size. Among the VietGAP farmers, 62% responded that they needed labor to sort and/or pack their products before delivering to buyers (Table 12.3).

In contrast, conventional farmers who do not have a contractual relationship with middlemen and/or wholesalers, decide the types of vegetables for each season based on their own experience (94%). Only 20% of the conventional farmers in the survey responded that they had established some kind of contract[4] with buyers but not for their entire cultivated areas (Table 12.2). The rest of the conventional farmers responded that they sold their vegetables to traditional middlemen and/or wholesalers in the traditional market. The corollary is that most of the conventional farmers (83%) depend heavily on middlemen and have to accept precarious prices (Table 12.2).

Regarding technical support, 65% of the VietGAP farmers consult the VietGAP buyers for advice on vegetable cultivation, whereas 95% of conventional farmers obtain technical information via informal personal relations (Table 12.4). Interestingly, no VietGAP farmers in our sample rely on agricultural officers (from the local government) for technical support, despite the fact that the VietGAP is a public program.

In sum, the relationship between the VietGAP farmers and their buyers is maintained during the entire production period, while most conventional farmers only start to bargain with middlemen and/or wholesalers at the time of harvest. As a result, the VietGAP buyers not only play the role of collectors but also govern the entire production process of the VietGAP vegetables. This innovative form of partnership is the key element for a higher level of food safety and traceability in this supply chain.

The more intensive use of labor inputs and compliance with strict regulations, as well as requirements from buyers, result in higher revenues for the VietGAP farmers, of which 91% reported revenues exceeding those of conventional farmers. Half of the VietGAP farmers could increase their revenues by more than 31% (Table 12.5). This implies that higher revenues, the most important motive

Table 12.4 Technical advice for vegetable cultivation

	Technical advice for vegetable cultivation				
	The VietGAP buyers	Cooperative*	Agricultural officers	Relatives	Others
The VietGAP farmers (n = 34)	22 (65%)	4 (11%)	0 (0%)	3 (9%)	5 (15%)
Conventional farmers (n = 41)	0 (0%)	0 (0%)	2 (5%)	11 (27%)	28** (68%)

Source: Authors' survey.
Note: The percentages are in parentheses.
* Cooperative; farmers' production group based on cooperative law in Vietnam.
** The other sources for technical advice that these 28 farmers reported include the fertilizer and pesticide suppliers (18 cases), farmers' own experience (14 cases), the internet and media (two cases), and wholesalers/middlemen (two cases).

Table 12.5 Revenue change among the VietGAP households/farmers

Revenue change compared to conventional cultivation	No. of farmers	Percentage (%)
Decrease	1	3
Almost no changes	2	6
Increase no more than 30%	14	41
Increase more than 31%	17	50
Total	34	100

Source: Authors' survey.

for farmers to participate in the VietGAP program, have been achieved in a short period, which is likely to speed up the expansion of the VietGAP cultivation.

How do conventional farmers switch to become the VietGAP producers?

To explore factors motivating farmers to adopt the VietGAP in their cultivation, the way in which they access this program was examined. In addition, based on their personal view, both groups of farmers were asked to choose the most important condition that facilitates the becoming of a VietGAP farmer.

Even though the VietGAP program was introduced by the government, only 9% of the VietGAP farmers obtained information regarding the program directly from local agricultural extension officers. Fifty-nine percent of respondents were introduced to the program by the VietGAP buyers, who claimed to play a decisive role in the transition from conventional to the VietGAP farming practices among farmers. Most of the VietGAP buyers in our study site used to be traditional middlemen and farmers, supplying vegetables to supermarkets and agro-food companies even before participating in the VietGAP program. Similarly, the majority (56%) of the farmers learned the VietGAP technique from these buyers (Figure 12.3). To be a VietGAP farmer, farmers must attend training courses organized by the government. In spite of these official training courses, the VietGAP farmers depended on their buyers for technical support, which strengthened the role of these buyers in the VietGAP implementation.

The data from the VietGAP sample confirm what they perceive to be the necessary conditions to become a VietGAP farmer. Sixty-two percent responded that the most important requirement to participate in the VietGAP vegetable

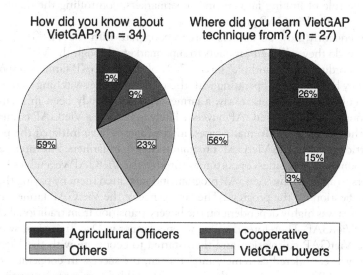

Figure 12.3 Access to the VietGAP Cultivation.

Source: Authors' survey.

Table 12.6 Farmers' perceptions on the necessary conditions to become a VietGAP farmer

	The most important condition to become a VietGAP farmer				
	Attending training courses	Having technical knowledge on the VietGAP farming	Investments in high-tech facilities	Having good relations with the VietGAP buyers	Others (no. idea)
The VietGAP farmers	2	10	1	21	0
(n = 34)	(6%)	(29%)	(3%)	(62%)	(0%)
Conventional farmers	1	1	6	10	23
(n = 41)	(2%)	(2%)	(15%)	(25%)	(56%)

Source: Authors' survey.
Note: The percentages are in parentheses.

production was having good relationships with the VietGAP buyers rather than attending training courses or being equipped with technical knowledge on the VietGAP farming (Table 12.6).

To explain their decision to switch, all the VietGAP farmers admitted that they decided to participate in the program with the expectation of modern market access and premium earnings from the VietGAP products. The motivation to apply to be a member of the VietGAP is accompanied by a change in market orientation. As individual farmers can hardly satisfy the large quantity requirements of supermarkets, they tend to access a supermarket indirectly by selling vegetables through the VietGAP buyers. In other words, the VietGAP buyers play the role of linking farmers and supermarkets, controlling the quality and quantity of the VietGAP products.

Subsequently, how do farmers start to cooperate with the VietGAP buyers, and how do they sell their products to supermarkets through the VietGAP buyers? According to the authors' survey, 76% of the VietGAP farmers responded that they had started cooperating with these buyers before switching to the Viet-GAP cultivation. That is to say, a farmer who has already been in a business relationship with the VietGAP buyers is likely to become a VietGAP farmer, not the other way around. In many cases, intermediaries have initiated the process of participating in the VietGAP program rather than farmers. An intermediary who found a new business opportunity through the VietGAP vegetables invited farmers to apply to the VietGAP program and supported them by paying the cost of certification for the program. The emergence of the VietGAP farmers in the study area was highly dependent on the buyers' transition from traditional chains to the VietGAP. Therefore, the case when a farmer in Duc Trong first switched to the VietGAP cultivation and then started to cooperate with the VietGAP buyers was considered extremely rare. In sum, the success in cooperating with the VietGAP buyers (supermarket suppliers) enabled farmers to enter the new vegetable distribution channel.

Governing the production of safe vegetables by the VietGAP buyers

Accompanied by the promotion of the VietGAP standards by the government and the increased demands of "safe vegetables" in urban areas, the VietGAP buyers have emerged in the production areas, stimulating the linkage between farmers and supermarkets. They have contributed to the establishment of a new distribution channel for the VietGAP vegetables. They have organized the Viet-GAP farmers to satisfy the supermarkets' requirements of large quantities, reliable quality, and a wide variety of vegetables. They govern the whole process of vegetable production through formal contracts with the VietGAP farmers. Conversely, farmers who have difficulties in bargaining directly with supermarkets due to small-scale production need to cooperate with the VietGAP buyers to access the modern supply chain.

To receive a VietGAP certificate, each farm must satisfy the 65 criteria of the VietGAP standards, which will be certified by a third party. This imposes high costs on farmers between 500 and 800 USD per hectare for two years. The VietGAP buyers in Duc Trong usually support farmers by paying these costs or assisting them in getting government subsidies for certification. As a result, most of the VietGAP certificates in Duc Trong are often issued in groups rather than to individual farmers. In this way, the VietGAP buyers own certificates as representatives of certified farmers, coordinating a number of member farmers and their certified plots. A collective certificate, in turn, pushes member farmers to comply with quality management. If a member is caught by a third certification party or the authorities for not complying, the collective certificate will be revoked, and all members cannot sell their products as the VietGAP vegetables. Therefore, the collective VietGAP certification strengthens the bond among members, including the VietGAP buyers, based on responsibilities and benefits.

According to the interviews with the VietGAP buyers in Duc Trong, after receiving production orders from supermarkets, they contact their member farmers to discuss the production schedule, providing technical services such as seeds and guidance on cultivation. In addition, they conduct internal inspections to ensure compliance of the member farmers with the VietGAP standards. Furthermore, they share necessary information related to the requirements of supermarkets on the quality of fresh vegetables.

Most of the interviewed VietGAP buyers in Duc Trong were originally vegetable farmers themselves before the implementation of the VietGAP and were innovative with relatively high educational levels. While carrying out their own cultivation, they also collected their neighbors' vegetables for redistribution as local middlemen. In the context of rural Vietnam, it is not extraordinary to work both as a farmer and a middleman. Their businesses as middlemen depended strongly on their social networks that were loosely organized for a long time, such as relatives, neighbors, and friends. This is the reason for the 76% pre-acquaintance between the VietGAP farmers and their buyers mentioned above. Using these social networks, the VietGAP buyers have reorganized vegetable farmers in two different ways.

The case of vertical integration

Tien Huy Cooperative was established in 2014 by Mr. Huy, the representative, and the seven VietGAP farmers. He used to be a medium-scale middleman in the vegetable supply chain. Mr. Huy began trading "conventional" vegetables with several supermarkets and vegetable processing companies in 2005. After recognizing that "safe vegetables" could bring about a good business opportunity, he persuaded his farmer clients to join the VietGAP program, supporting them by paying the cost of certification. In 2009, he received the VietGAP certification for 0.9 hectares and became a VietGAP buyer.[5] In the subsequent year, he succeeded in persuading 11 farmers in the same district to join the VietGAP program and supported them by paying the cost of certification. In this process, he could contract with the farmers on the exclusive right to purchase their VietGAP vegetables from the certified land (10 hectares). At the time of the field survey in 2019, the cooperative had 41 member farmers with 27.5 hectares of certified land. It produced about 10 different types of vegetables with an output of approximately 1,500 tons per year. Furthermore, it employed 13 regular staff members and about 30 part-time workers during the peak season for quality checks, as well as resorting and repacking. The technical staff visit their member farmers once a week for internal inspection. In fact, as Mr. Huy was a medium-scale middleman, it was easy for him to become a VietGAP buyer. Traditional large-scale middlemen and/or wholesalers in the Duc Trong district, who dealt with huge amounts of only a few types of vegetables, did not want to trade the VietGAP vegetables due to inertia and difficulties in quality control. However, the establishment of the cooperative made it likely for them to receive government subsidies for the cost of certification. The real operation of the cooperative was similar to that of private companies. The source of profits for the cooperative was the commission for handling vegetables from the member farmers. The small-/medium-scale middlemen could have an opportunity to trade directly with supermarkets through vertical integration.

Horizontal integration

The Hoi Toan Cooperative was established in 2015 with 10 member farmers to access modern distribution channels. The cooperative received the VietGAP certification for 11 hectares in the same year, with the cost of certification being subsidized by the government. Mr. Hoi, the leader of the cooperative, established the cooperative with some innovative farmers within 2 km from his village. In 2017, they started supplying the VietGAP vegetables directly to the VinEco Da Lat Platform, a purchasing center of the VinMart supermarket chain. At the time of the field survey in 2019, the number of members had increased to 14 farmers with 17 hectares of certified land. Before participating in the VietGAP program, all members were vegetable growers and sold their products to local middlemen. Despite the disadvantages of small-scale farming without connections with the VietGAP buyers in the beginning, they successfully accessed the new distribution channel through self-organization. The operation of the cooperative was based on the principle of equal rights according to shares. Extra payments were made for labor contributions such as sorting, packing, and transportation. The small farmers were able to bargain directly with supermarkets through the horizontal integration of the cooperative model.

Conclusion

The implementation of the VietGAP program in Vietnam has changed the relationship between the VietGAP farmers and their buyers into a contractual one for the entire production period. As a result, the VietGAP buyers not only played the role of collectors but also governed the whole process of the Viet-GAP vegetable production. The relationship with the VietGAP buyers played a critical role in transforming from a conventional to a VietGAP farmer. In other words, a farmer who already had connections with the VietGAP buyers eventually became a VietGAP farmer, not that a farmer first converted to the VietGAP cultivation and then started to cooperate with the VietGAP buyers. The success of cooperating with the VietGAP buyers (supermarket suppliers) makes it possible for farmers to enter the new vegetable distribution channel.

The results of this study revealed that the VietGAP implementation has had an impact on the reorganization of stakeholders in the local supply chain, leading to the development of a new distribution channel for the VietGAP vegetables, in addition to the traditional supply chain. It has also strengthened the linkages among stakeholders in the new distribution channel through contractual relationships. Our results highlight that the VietGAP buyers have emerged as the main actors in the newly established distribution channel, playing a key role in creating linkages between vegetable farmers and supermarkets.

Two cooperative models of reorganizing farmers by the VietGAP buyers were observed. First, middlemen, who had business relationships with supermarkets prior to the implementation of the VietGAP, were motivated to organize farmers so as to meet their clients' (supermarkets) demand for the VietGAP vegetables. To ensure the provision of large and stable amounts of the VietGAP vegetables, they reorganized loosely connected farmers within their geographical areas into contracted groups for the VietGAP cultivation. This relationship is vertical, where the buyer governs the entire production process. The other case is when farmers organized themselves and established a VietGAP vegetable production cooperative. Vegetable farmers willing to adopt the VietGAP without social relations with the VietGAP buyers and/or supermarkets so as to access the modern distribution channel established a cooperative to expand the VietGAP cultivation areas to satisfy the requirements of supermarkets. With the increased production capacities, they could sell their products directly to supermarkets. The relationship among these members is horizontal, and quality management is performed by mutual inspections.

In sum, the implementation of the VietGAP, an agricultural product certification system in Vietnam, has succeeded in establishing a new distribution channel by reorganizing farmers into contractual relationships. With this new supply chain, a higher level of food safety and traceability is achieved due to governance over the entire production process by the buyers, while farmers benefit from a more modern mode of food production and distribution.

Acknowledgments

We thank JSPS KAKENHI Grant Number 18K01140 for funding part of this research. In addition, we would like to express our gratitude to farmers in the Duc Trong district, who patiently received our long interviews and provided us with valuable data for our research. Finally, we would like to express our appreciation to Editage (www.editage.jp) for careful English language editing.

Notes

1 Source of data: Department of Agricultural and Rural Development, Lam Dong Province, 2016.
2 A VietGAP farmer in this research is defined as a farmer who grows vegetables in at least one plot certified through the VietGAP program. Some VietGAP farmers in the study area grow vegetables separately according to both the VietGAP standards and the conventional way in different plots.
3 A VietGAP buyer in this research is defined as an intermediary who collects vegetables from the VietGAP farmers and supplies them to modern food retail facilities such as supermarkets. At the time of the field survey, eight VietGAP buyers were operating in the Duc Trong district. A few VietGAP buyers deal with both the VietGAP vegetables and conventional vegetables, with different distribution channels. The extra VietGAP vegetables were also supplied as conventional vegetables through traditional supply chains.
4 These types of contracts with traditional middlemen and/or wholesalers are exchanged at the time of harvest, not before the cultivating season.
5 To deal with the VietGAP vegetables, a buyer must have the VietGAP certificate, which is often issued in groups as explained above. However, in 2009, only a few farmers had the VietGAP certification; therefore, he had to apply for the VietGAP certification for his own land.

References

Abebe, G. K., J. Bijman, and A. Royer. 2016. Are Middlemen Facilitators or Barriers to Improve Smallholders' Welfare in Rural Economies? Empirical Evidence from Ethiopia. *Journal of Rural Studies* 43: 203–13.
Cadilhon, J.-J., A. P. Fearne, P. Moustier, and N. D. Poole. 2003. Modelling Vegetable Marketing Systems in South East Asia: Phenomenological Insights from Vietnam. *Supply Chain Management: An International Journal* 8 (5): 427–41.
Cadilhon, J.-J., P. Moustier, N. D. Poole, T. G. T. Phan, and A. P. Fearne. 2006. Traditional vs. Modern Food Systems? Insights from Vegetable Supply Chains to Ho Chi Minh City (Vietnam). *Development Policy Review* 24 (1): 31–49.
van Duijn, A. P., R. Beukers, and W. Van der Pijl. 2012. The Vietnamese Seafood Sector: A Value Chain Analysis. CBI/LEI, part of Wageningen UR.
Hoang, H. G. 2018. Farmers' Responses to VietGAP: A Case Study of a Policy Mechanism for Transforming the Traditional Agri-Food System in Vietnam: A Dissertation Presented in Partial Fulfilment of the Requirements for the Degree of Doctor of Philosophy in Agricultural Systems and Environment at Massey University, Palmerston North, New Zealand.
Hu, K., J. Liu, B. Li, L. Liu, S. M. T. Gharibzahedi, Y. Su, Y. Jiang, J. Tan, Y. Wang, and Y. Guo. 2019. Global Research Trends in Food Safety in Agriculture and Industry from 1991 to 2018: A Data-Driven Analysis. *Trends in Food Science & Technology* 85 (March): 262–76.

Ministry of Agriculture and Rural Development. 2008. Decision No. 99/2008/QD-BNN.

Moustier, P., T. G. T. Phan, T. A. Dao, T. B. Vu, and T. T. L. Nguyen. 2010. The Role of Farmer Organizations in Supplying Supermarkets with Quality Food in Vietnam. *Food Policy* 35 (1): 69–78.

Nabeshima, K., E. Michida, H. N. Vu, and A. Suzuki. 2015. Emergence of Asian GAPs and Its Relationship to Global GAP. 507. IDE Discussion Paper. Inst. of Developing Economies, Japan External Trade Organization.

National Assembly of Vietnam. 2010. Law No. 55/2010/QH12 on Food Safety.

Ngo, H. M., Q. H. Vu, R. Liu, M. Moritaka, and S. Fukuda. 2019. Challenges for the Development of Safe Vegetables in Vietnam: An Insight into the Supply Chains in Hanoi City. *Journal of Faculty of Agriculture Kyushu University* 64 (2): 355–65.

Nguyen, M. H. D., M. Demont, E. J. van Loo, A. de Guia, P. Rutsaert, H. T. Tran, and W. Verbeke. 2018. What Is the Value of Sustainably-Produced Rice? Consumer Evidence from Experimental Auctions in Vietnam. *Food Policy* 79 (August): 283–96.

Nguyen, M. H. D., P. Rutsaert, E. J. van Loo, and W. Verbeke. 2017. Consumers' Familiarity with and Attitudes towards Food Quality Certifications for Rice and Vegetables in Vietnam. *Food Control* 82 (December): 74–82.

Nguyen, D. H., C. T. Tran, V. H. Nguyen, T. L. Nguyen, V. M. Dang, D. T. Trinh, T. L. N. Huynh, T. P. Nguyen, and T. S. Thai. 1999. Impact of Agro-Chemical Use on Productivity and Health in Vietnam. EEPSEA Research Report Series. Economy and Environment Program for Southeast Asia. https://idl-bnc-idrc.dspacedirect.org/bit-stream/handle/10625/22624/112823.pdf?sequence=5

Nicetic, O., E. van de Fliert, V. C. Ho, M. Vo, and C. Le. 2010. Good Agricultural Practice (GAP) as a Vehicle for Transformation to Sustainable Citrus Production in the Mekong Delta of Vietnam. *9th European IFSA Symposium, Vienna Austria*, July 4–7, 2010. University of Natural Resources and Applied Life Sciences, Vienna. http://ifsa.boku.ac.at/cms/fileadmin/Proceeding2010/2010_WS4.4_Nicetic.pdf

Niewöhner, J., A. Bruns, P. Hostert, T. Krueger, J. Ø. Nielsen, H. Haberl, C. Lauk, J. Lutz, and D. Müller, eds. 2016. *Land Use Competition: Ecological, Economic and Social Perspectives*. Cham: Springer International Publishing.

Ouma, S. 2010. Global Standards, Local Realities: Private Agrifood Governance and the Restructuring of the Kenyan Horticulture Industry. *Economic Geography* 86 (2): 197–222.

Pham, V. H., A. P. J. Mol, and P. J. M. Oosterveer. 2009. Market Governance for Safe Food in Developing Countries: The Case of Low-Pesticide Vegetables in Vietnam. *Journal of Environmental Management* 91 (2): 380–88.

Reardon, T. 2015. The Hidden Middle: The Quiet Revolution in the Midstream of Agrifood Value Chains in Developing Countries. *Oxford Review of Economic Policy* 31 (1): 45–63.

Reardon, T., K. Z. Chen, B. Minten, L. Adriano, A. D. The, J. Wang, and S. D. Gupta. 2014. The Quiet Revolution in Asia's Rice Value. *Annals of the New York Academy of Sciences* 1331 (1): 106–18.

Reardon, T., R. Echeverria, J. Berdegué, B. Minten, S. Liverpool-Tasie, D. Tschirley, and D. Zilberman. 2019. Rapid Transformation of Food Systems in Developing Regions: Highlighting the Role of Agricultural Research & Innovations. *Agricultural Systems* 172: 47–59.

Reardon, T., and C. P. Timmer. 2014. Five Inter-Linked Transformations in the Asian Agrifood Economy: Food Security Implications. *Global Food Security* 3 (2): 108–17.

Reardon, T., C. P. Timmer, C. B. Barrett, and J. Berdegué. 2003. The Rise of Supermarkets in Africa, Asia, and Latin America. *American Journal of Agricultural Economics* 85 (5): 1140–46.

Tennent, R., and S. Lockie. 2013. Private Food Standards, Trade and Institutions in Vietnam. *Journal of Asian Public Policy* 6 (2): 163–77.

Tran, N., C. Bailey, N. Wilson, and M. Phillips. 2013. Governance of Global Value Chains in Response to Food Safety and Certification Standards: The Case of Shrimp from Vietnam. *World Development* 45 (May): 325–36.

Tran, D., and D. Goto. 2019. Impacts of Sustainability Certification on Farm Income: Evidence from Small-Scale Specialty Green Tea Farmers in Vietnam. *Food Policy* 83 (February): 70–82.

Wang, H., P. Moustier, and T. T. L. Nguyen. 2014. Economic Impact of Direct Marketing and Contracts: The Case of Safe Vegetable Chains in Northern Vietnam. *Food Policy* 47 (August): 13–23.

Wang, H., P. Moustier, T. T. L. Nguyen, and T. H. T. Pham. 2012. Quality Control of Safe Vegetables by Collective Action in Hanoi, Vietnam. *Procedia Economics and Finance* 2: 344–52.

Wertheim-Heck, S. C. O., S. Vellema, and G. Spaargaren. 2015. Food Safety and Urban Food Markets in Vietnam: The Need for Flexible and Customized Retail Modernization Policies. *Food Policy* 54 (July): 95–106.

13 Relocalizing Food Systems for Everyone, Everywhere? Reflections on Walloon Initiatives (Belgium)

Antonia Bousbaine and Serge Schmitz

Introduction

Since 1962, the main concern of European agricultural policy has been to pro-duce more and cheaper food, which allows, inter alia, European consumers to spend money on commercial goods and services other than food. Meanwhile, the food issue has been part of many debates at both the consumer and polit-ical actor levels. Whether at the local, regional or national level, food issues have been a global topic of discussion (Renting et al. 2003; Goodman 2004; Van der Ploeg and Renting 2004; Maye et al. 2007; Chiffoleau 2008; Maréchal 2008; Wiskerke 2009; Brand 2015; Billion 2018). There is a growing aware-ness of defects in the conventional food system resulting from chemical inputs, industrial food processing processes, the internationalization of markets and the lack of traceability concerning the origin, composition and processes of food production. These concerns relate to the quality of the food, the impacts of its production on the environment and human health, and the dominance of both industrial processing and international supply chains by a small number of large commercial concerns.

For the past few decades, agricultural areas around cities have regained an interest in directly feeding cities. Since the Second World War, the relationship between cities and agriculture has been neglected in most West European coun-tries. The agricultural lands surrounding cities were viewed as potential building areas for housing and diverse productive economic activities or in other places as an attractive territory for wealthy peoples' residence and for recreation. When farming was sufficiently competitive in international markets, it disconnected from its regional embeddedness and focused on producing standard food for the world market. Conversely, the development of agriculture in less favourable physical environments or more peripheral locations has experienced the aban-donment of farming activities. Recently, in many places in the Global North, especially former industrial cities (such as Detroit, Manchester, Milan and Liège), citizens and politicians have reconsidered their links to food and their agricultural belts. This situation can lead to a profound rural transformation of the countryside and agricultural practices near urbanized areas and a renewed relationship and governance between towns and the countryside.

DOI: 10.4324/9781003110095-16

This chapter aims to document the renewed interest in local food and regional agriculture and note the rural transformations required by this evolution. The work was conducted over a period of four years; it followed the emergence of innovative agro-food systems, which constitute the belts of local food production around Liège and Charleroi, Belgium (Figure 13.1). In addition to these belts, numerous local food projects have also been identified. We will scrutinize their conditions of emergence and the numerous economic, cultural and political constraints that have been identified. The following questions therefore arise: can we attain a better understanding of what these initiatives are, and can these initiatives provide adequate quality food to all the populations concerned? Indeed, reconstructing a food link between the city and farmers from the region should require investment on both sides and probably external or political help to create new and more sustainable relationships between actors and places.

Wallonia, one of the three federal regions of Belgium (Figure 13.1), is not to be outdone. As elsewhere, the food system has been substantially considered. A food strategy was recently implemented by the office of the Minister of Sustainable Development and Ecological Transition in November 2018. Initially, sustainable food forums were organized by both consumers and actors in the food system. Little by little, therefore, the issue of relocalizing the food system has arisen in Wallonia.

This part of Belgium includes 3.65 million inhabitants with an average density of 215 inhabitants/km² (Iweps 2020), making this region one of the most densely

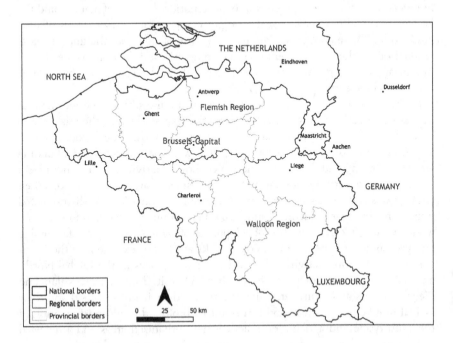

Figure 13.1 Location map.

Source: Laplec.

populated in Europe. Rich in a past dominated by steel and coal mining activities, Wallonia has long relied on these activities, which made it a real economic power at the end of the 19th century. The last coal mine ceased its activity in 1980, and the steel industry drastically reduced the number of plants, leaving behind many brownfield sites. These sites have sometimes been reused by innovative urban agricultural practices. Mines and industries accounted for most of the labour force of the country (including horses and dogs) and deserved the full attention of financial and political arenas, weakening the agricultural sector (Schmitz 2013). The definition of the Belgian rural world was largely influenced by industrial and urban development. Although the countryside had previously been the cradle of an abundant workforce (van Hecke et al. 2010), very quickly, it was monopolized by city dwellers who found a way of life that diverges from the city and involved access to property (Halleux et al. 2002). In this sense, "the persistence of a rural world" in Belgium has arisen in view of the drastic loss of agricultural land and farms (Schmitz 2002). Over 30 years, Belgium has experienced a reduction of 70% of its farms (Bousbaine 2020).

This research focused on 18 municipalities from the metropolitan areas of Liège and Charleroi selected according to agri-urban projects. We conducted 33 interviews with various players in the field: local development agencies, local action groups (linked with the LEADER programme), mayors or deputy mayors for agriculture, provincial and European deputies, ministers (of agriculture and the economy), and officials of the Walloon agricultural unions (FWA and FUGEA). We also analysed a national newspaper (*Le Soir*) to list the initiatives and the defended values. The interviews focused on the agri-food issue, whether any importance was associated with agricultural activity, the initiatives enacted and the interest associated with the food issue. Finally, to compare the visions of these key players with the realities on the ground, we conducted phone and door-to-door surveys with a sample of 399 inhabitants and 98 farmers from the 18 municipalities. The survey addressed the perception of agriculture by farmers and non-farmers, the understanding of the agricultural contributions, the advantages and inconveniences of maintaining agriculture in a commune, the initiatives and the orientations necessary to bring the agricultural world and the inhabitants closer, and finally the supply mode for fruits and vegetables.

Charleroi (291,000 inhabitants) and Liège (501,000 inhabitants) are the two largest Walloon conurbations. After the Second World War, these two cities experienced declines in their steel, glass and coal mining activities. The agriculture developed in the Loess Plateau around Charleroi is mainly oriented to field crops: sugar beet, cereals and potatoes. In Liège, dairy farming is dominant in the hilly landscape, but there are also field crops on the Loess deposits, and fruit crops have decreased significantly since 1950 (Bousbaine 2020). In these cities, the food issue has been integrated into the consumer sphere and then joined by political authorities. In Liège, food relocation initiatives are plentiful and constantly innovating. They relate notably to the food land belt concept, which was established in 2013. In Charleroi, the trend is less palpable, despite the establishment of a food land belt that is largely inspired by Liège.

A reshaped demand for fresh food from the city to the countryside

In Wallonia, numerous newspaper articles have stated the desire to relocate food within spaces, which involves the development of short food circuits (Bousbaine 2020). These short food circuits aim to both reduce the number of intermediaries and the distance between producers and consumers (O'Hara and Stagl 2001; Praly et al. 2014); the same observations have been documented in numerous countries by an extensive literature (Maréchal 2008; Wiskerke 2009).

The health crisis as triggering factor

The renewal of interest for local food and regional agriculture has become central since the health crises of the 1990s, including spongiform encephalopathy, Belgian PCB and the dioxin crisis (1999), or, recently, the Fipronil eggs contamination (2018). Food questions have been re-examined by authorities as well as by consumers. These issues call into question the quality and the origin of their food supply and the dominant agro-industrial system that has presented severe dysfunctionalities (Bousbaine 2020). These problems have been linked with the organization of territories, the loss of jobs relating to farming (Ericksen 2008; Rastoin and Ghersi 2010) and with environmental issues relating to the massive use of chemical fertilizers, the over-exploitation of land and other resources, soil erosion, damage to biodiversity and global warming (Theys and Vidalenc 2013). Finally, at the social level, this food system has distanced producers from consumers (Prigent-Simonin and Hérault-Fournier 2012) by "disconnecting agriculture from its territories" (Vachon 1991; van der Ploeg and Renting 2004; Wiskerke 2009). This has been generating a division between places of production, processing and marketing, resulting in multiple food landscapes (Ilbery et al. 2000). Indeed, these food market systems have generated food insecurity regarding both quantity and quality (FAO 2003; Maxwell and Slater 2003; Janin and Dury 2012). While several hundred farms have been disappearing, the remaining farmers experience huge difficulties surviving in the international market; therefore, it seems urgent to address the question of the preservation of agricultural lands and to the food issues that result from it. Finally, food quality is an issue that is upsetting political authorities who have become aware of its impact on public health. However, a lengthy period will be experienced before this idea to feed the city, mostly through the production of urban and suburban agriculture, can be implemented.

Food purchasing groups in Charleroi and Liège

One of the first initiatives noted in Liège city was the Barricade joint purchasing group, which was formed in June 1999. After long reflections and discussions within the cultural centre of Barricade, a group of citizens positioned the ways they eat and supply. These discussions were conducted with a community and a cooperative farm, Ferme du Hayon, which claims to be peasant agriculture producing agroecologically (Louviaux 2011).

Based on a desire to find healthy, quality food and to support local farmers, this project brings together 20 families from the neighbourhood and a few other people interested in the process. This food purchasing group is a form of short circuit that encourages bypassing the dominant system.

This group has been emulated by other districts of the city. The centre of Liège no longer has any farm capable of feeding the population, and it is therefore on the outskirts that the joint purchasing groups find these local producers. These include the "Ferme Larock", which aims to feed the local population using the biodynamic method, and the "Ferme de l'Arbre", which cultivates organically.

The interviews showed that these initiatives are running out of steam and rely often on a few members. Moreover, these initiatives are still struggling to get out of their field of action, namely, the local domain, to be part of a more global approach allowing greater visibility of their commitment and to break out of their confinement to a niche "logic" (Verhaegen 2011).

In Charleroi, there are very few food purchasing groups. We discovered three groups: Gilly was initiated in 2014 by a local association; Mont-sur-Marchiennes was created in 2013 in a spirit of sustainable and social development; for a year, the creation of the Coopéco cooperative brought together 400 members in the centre of Charleroi. The group in Mont-sur-Marchiennes calls on an organic producer, the Ferme Maustitchi, located in Fontaine-l'Évêque, which has been supplying the population for many years via short circuits.

In addition to these food purchasing groups, we have noted innovative initiatives from consumer and producer cooperatives in Charleroi and Liège. In the centre of Charleroi, a citizen's cooperative wants to provide healthy and accessible food to many people. Based on the evidence of the unsustainability of the principal food system, this cooperative bypasses mass distribution by establishing a store that aims to be ethical and in which local products are sold by involving each co-operator. This "participative store" directly involves the cooperative members who become owners, warehouse workers and managers. Among the co-operators, the majority come from the middle class. They are essentially inhabitants living in the city centre.

In Liège, cooperatives are well established in the city landscape. The store "Les petits producteurs", which emanates from the food land belt (see below), has been installed in the centre of Liège since 2016. This store is a social cooperative that promotes access to local and healthy products for all and supports producers from the province. This store works especially with local NGOs such as La Bourrache and the Coopérative Ardente, a local organic and fair-trade grocery store. Building on this success, with more than 1000 customers a week, a second store opened a year later and a third store opened in December 2018. In the neighbouring municipality of Saint Nicolas, the Coopérative Ardente delivers more than 800 food and household products directly to customers or at collection points. Another cooperative "Point Ferme", whose slogan is "for sustainable, local and equitable agriculture", was created in 2011 to join 30 producers in 2020 to supply more than 500 baskets per week at 50 relay points. All these initiatives have been boosted by the COVID-19 crisis and appeared as an "alternative" to the mass agro-industrial food system.

The food land belts

Nonetheless, the more challenging and promising initiatives are the establishment of food land belts because they aim to connect numerous actors from the food market and food processing system to ensure relative food independence. They should have social, economic and environmental impacts by maintaining agriculture in the suburbs and delivering fresh healthy food in cities.

This concept was born in 2013 in Liège, where for some years, a reflection on the food system occurred. A coalition of heterogeneous actors reflected on the transition aiming to feed a territory: the province of Liège. The food land belt, "Ceinture Aliment Terre" in French, was formed informally after discussions between social entrepreneurs and members of "Liège in Transition" during the conference "Innovative Alliance for Sustainable Development" in 2012. Based on workshops and round tables that established the ideas of this project of relocation, notably the decarbonization and greening of food in the province, the participants reflected on the alternative food systems. The implementation will benefit from the rise of both food purchasing groups and cooperatives and companies with a socio-economic orientation. Isolated initiatives have given rise to the idea of establishing a network in which existing alternative actions converge towards this desire for transition within the territory in terms of food. As such, the Liège Food Land Belt is a pioneer in terms of networking actors who are opting for a change in the food process and market systems. This belt links the city and the surrounding countryside, establishing an economic and food link and contributing to the preservation of agricultural land against urban sprawl (Alexandre and Génin 2014).

The question asked by the initiators of the project was as follows: "How can we manage, within 25 to 30 years, to ensure that most of the food consumed in the province of Liège (1.1 million inhabitants) is produced locally, in a sustainable and fair manner?" A reflexive and participative dynamic involving consumers has been established in order to highlight local food linked to the potential of the territory, thus connecting the city to its nourishing countryside. The participants reflected on a network of actors from short circuits in order to co-construct projects combining producers and consumers. Consumers act knowingly and participate directly in their food choices; they are at the heart of the project by contributing to the initiatives to be put in place, in which they become fully "citizens" and act for the benefit of the community (Baguette 2015).

The specificity of the Liège Food Land Belt resides in the involvement of all actors from the territory without distinction, including civil society, local communities, and the private and economic sectors, in a bottom-up approach that starts with the territory and its actors. The objective is to join stakeholders by identifying the citizens' needs and the resources available within the territory. Faced with the shortage of local food producers in short circuits and the high demand from consumers, the belt has set the objective of supporting and training project leaders. It aims to be the missing "link" to unite all the driving forces. It structures the projects that are being established by guiding and supporting them, not in a sectoral way but in a territorial manner. The final challenges focus on the collective reappropriation of the food chain, ensuring that each inhabitant can have access to healthy and sustainable food. Although the Liège Food Land Belt is the first to date in Wallonia, one of its objectives is to "spread" throughout the rest of Wallonia and beyond.

For the designers, the systems to be established must converge with the agricultural reality that exists in the province of Liège, where agriculture is struggling to resist the various crises that have hit the agricultural world, especially the dairy crisis. Additionally, to meet the demands for local products, agricultural systems must transition towards more sustainable, more ecological systems. Nonetheless, a large proportion of farmers is not interested in being part of this project because they have invested years and much money to exist in the conventional system. Feeding local populations is then seen as a totally new business.

In Charleroi, certain players in the voluntary sector have been seeking to raise awareness of the merits of food relocation among the population. Considering the lack of consideration for the food issue by both local public authorities and most of the population, some key players in short food circuits responded to a call from the Agency for Enterprise and Innovation from the Walloon government in 2016. The food land belt project of Charleroi was initiated. The initiators are a social enterprise that claims the importance of food in cities; the relocation of this food, which contributes to urban metabolism; and Biowallonie, which promotes organic farming in Wallonia.

The main vision of this food land belt is as follows: to thoroughly review the existing agro-food system in Charleroi Métropole, on a local scale, by co-constructing projects that involve different actors (farmers, associations, scholars, traders, artisans, local elected officials and citizens) of the territory to bring together the expectations of each other in order to set up food systems that feed the populations through local products, at affordable prices for all, based on organic, peasant and local agriculture. The initiators of this project opt to draw together producers, traders and consumers in a spirit of cooperation and reciprocity so that each of the actors involved is in a more comfortable economic situation. This belt delimits supply zones that constitute the rural zones where agriculture subsists. The catchment area represents the consumption basin of Charleroi city and the periurban municipalities. The goal is therefore to match demand with supply, a problem that the initiators of this belt are focusing on. After a long implementation process, two meetings were held in August and September 2017 to present the aims and the involved actors. In Charleroi, only organic farmers are involved. The question therefore arises of how to feed a conurbation such as Charleroi with very few organic producers.

The purposes of the belt are as follows: to improve and increase the supply of these products via existing suppliers and producers who are newly established or who wish to convert to organic farming; to develop projects in the food processing, logistics and distribution fields; to develop collective, shared vegetable gardens in the city; to improve and increase the sale and access to local organic products in the city of Charleroi (and the surrounding area); to improve and increase supplies (and logistics), and to increase the awareness of citizens of healthy and local food.

The methodology adopted by the designers was first a question of diagnosing the demand and supply of local products in the conurbation in light of the resources available and then quantifying the needs of the actors considering the logistical issues. If some farmers seem interested in selling in short circuits, many people mentioned the difficulties existing in their distribution. An inventory listed approximately 20 organic farmers in the area.

In March 2018, a meeting helped identify the sales methods for organic products. Two solutions have been chosen: a cooperative store modelled on the model of "Les petits producteurs" in Liège and e-commerce.

In this stage, the food issue does not truly focus the attention of the mayors of the municipalities of Charleroi who must face major social, economic and urbanistic issues. If this project is gaining ground and is seeking to set the necessary markers in an area where almost everything must be constructed, the various interviews underlined that the food land belt is still at an embryonic stage and still has no real anchor within the population. Moreover, the lack of organic producers seems to constitute an obstacle that must be removed in the coming months.

This food land belt was strongly inspired by that of Liège whose main principles essentially overlap. It focuses on relocating food in the conurbation of Charleroi and thereby improving the sustainability of the food system. However, unlike in Liège, the participating producers must cultivate organically. Moreover, the populations in Charleroi are still far from possessing complete awareness of sustainable and local food, as evidenced by the rare projects identified.

The Food Land Belt of Charleroi Métropole faces many difficulties, including the socio-economic situation after the industrial collapse. However, this project operates over the long term. As one of the designers mentioned, there will be a long road to change people's minds and measure the importance of preserving local agriculture, both in terms of land use planning and its contribution to feeding local people.

The perception and the role of the farmers

Most of the initiatives listed in the two conurbations, including the food land belts, emerge from groups of consumers or NGOs who want to question the food system from a sustainable development point of view. Conventional farmers are under-represented in these food transitions. This is especially the case for the food land belt of Charleroi, where it was decided to restrict participation to organic farmers. This is also the case in Liège, where the more active farmers in the food land belt already practised direct selling. The role of the food purchasing groups, cooperatives and the food land belts is finally to create links between consumers who want to eat locally and responsibly and farmers who have already decided to produce and market differently. It seems difficult for conventional farmers to change both their minds and their ways of farming. There is a strong path dependency linking previous choices and investments.

In order to better identify the positions of farmers in both metropolitan areas, we surveyed 98 farmers and 399 inhabitants to explore the links between local agriculture and the city. The farmers declared that the relations between the farmers and the inhabitants are good in most cases. The principal inconvenience linked to their activity is related to the high price of agricultural land. The advantages retained by these respondents about these food relocation initiatives are to renew their links of proximity but also to participate in local economic life. In relation to the initiatives and orientations to be established, their choices were to organize activities on the farm, to establish a sort of charter of conviviality and to contribute to the education of youth in relation to healthy foodstuff. Sixty-three per cent of the farmers were disposed to feed the city, and these farmers supported the proposal of direct sales on farms and also opted for sales via intermediaries.

We note that if the relations with the inhabitants seem good, farmers suggest actions to enhance the cohabitation and the knowledge of agriculture. Considering food relocation, the orientation of conventional farming to the mass production market does not easily fit the production of food for the local population. Most farmers do not produce fruits or vegetables and produce sugar beets, wheat, potatoes, beef cows and milk that are transformed by industrial plants into products for distribution to the regional and international markets. Therefore, the relocation experiences are based on both types of producers: those who already have chosen to change their way of producing and marketing and those small farmers who have recently entered agriculture to produce differently. The challenge will be to know if these initiatives will be able to attract more farm operators, including those from conventional agriculture.

A similar remark can be made about consumers. If an increasing number of consumers are enticed and convinced by this food relocation, it requires that they change their habits and often accept investing time and slightly more money to purchase food. Another challenge will be to generalize this choice for local food and to make it affordable for the entire population.

The survey of 399 people from the 18 studied municipalities shows that 85% purchase fruits and vegetables in supermarkets. This tendency is lower for people older than 46 who retained the importance of the valorization and the organization of markets for local produce. Most inhabitants are committed to maintaining agriculture. More than 60% assess the relationship between farmers and the population as good even though only 43% have visited a farmer during the year. According to the survey, the presence of agricultural activity contributes to local economic life. However, the numbers continue to decline each year; moreover, the share of agriculture in the local economy seems to have substantially decreased. Another advantage is the creation of a closeness between consumers and producers and the purchase of local products, which is highlighted as a way to support farmers. Regarding the disadvantages, 45% of respondents stated smells. With the rise of environmental issues, rural residents are also being made aware of the harm caused by the use of pesticides and other chemical inputs, resulting in rejection of this type of agriculture. In addition, respondents highlighted the increase in land prices. While 63% of the surveyed inhabitants assess the relation between agriculture and residents as good, 11% of the inhabitants and 25% of the farmers declare that this relation is problematic. Indeed, there is a gap between the inhabitants who do not know too much about agriculture. In most cases, they are unable to assess the share of the territory occupied by agriculture or to give the names of the farmers. Reconnecting people with local food should also pass through a reconnection of people who produce and consume.

Discussion

In both studied conurbations, the food supply has been re-examined by an increasing share of the population for different reasons: healthy food, climate change, energy transition, solidarity with local farmers, tasty food, food independency, land and nature preservation. Even within the Walloon government, the Minister of the Environment has recently focused his attention on the food issue, which has been raised in several Walloon cities. This attempt to reconnect the city with its agriculture has been fully considered in the renovation plan of some Walloon cities launched by the Walloon government in 2015. A call for municipalities wishing to "participate in a global and integrated approach to sustainable development" evokes "nourishing zones". In this call for projects, the two priorities were centred on urban agriculture and citizen participation. The interest in urban agriculture echoes the initiatives noted in certain Walloon cities. Meanwhile, this recognition is also increasing in the rest of the world, such as in terms of land use planning (Van Veenhuizen 2006; Dubbeling and van Veenhuizen 2010); this renewed link with agriculture refers to the sustainability of cities (Schmitz 2008; Ba and Aubry 2011). Some of the projects analysed in the metropolitan areas of Liège and Charleroi capitalize on the possibilities of urban and periurban agriculture. How can agriculture improve the quality of the city? What forms and organizations already exist? How can local production be included in new neighbourhood projects and at the scale of the metropolitan area? The purpose of these projects should contribute to a better sustainability of the territories. A pending question is the limit of urban agriculture when the boundaries between cities and the countryside become more porous (Scheromm et al. 2019). How far can the food come from to be considered local? In the literature, according to the geographical context, the distance may be restricted to a few kilometres up to hundreds of kilometres or to national boundaries (Thompson et al. 2008; Johnson et al. 2013). In the two case studies, depending on the bearer of the initiative, the level of the project varies and may include the municipality, which is the more local political and administrative level in which urban areas may lack farmers and agricultural lands; metropolitan areas, which in Wallonia have no administrative statute; and the province.

Although the definitions of so-called urban agriculture are numerous (Mougeot 2000; Aubry and Pourias 2013; Poulot 2013; Mundler et al. 2014), its objectives receive a broad consensus: feed local populations to different degrees depending on the country. Its definition comes down to agriculture located in or around the city, whose products are mainly intended for the city and for which there is an alternative between agricultural and non-agricultural use of resources (land, water, labor work, etc.) (Moustier and Mbaye 1999). In a populated country such as Belgium, these questions regarding the use of resources are almost ubiquitous.

Therefore, countryside and agricultural lands should not only be protected but also need to find people to advocate for their preservation. Besides the precarious economic vitality of a few farms, there is a need to reallocate services to agricultural lands, including the feeding of the local population.

In the two studied conurbations, we noted certain initiatives emanating from citizens to reconsider their plates. Besides the creation of vegetable gardens, which were not discussed in this chapter, we noticed several initiatives that should impact the relation between cities and their countryside. The success of these initiatives depends at least on three factors: the awareness and involvement of citizens, the co-construction by actors from both the city and the countryside, and the possibility to produce locally despite the mass production dependency of most farmers. We noticed difficulties in both conurbations to match the new consumers' demand with local agricultural practices. In many cases, these projects face a lack of support from public authorities and the recurring problem of the affordability of agricultural land. In Wallonia, the price of agricultural land has substantially increased in recent decades, especially since land became a safe haven for investment during and after the financial crisis of 2008. The price has reached 40,000€, sometimes up to 100,000€, a hectare depending on the region (Bousbaine 2020) and constitutes a major obstacle to the establishment of new actors such as market gardeners who aim to feed the local populations. These market gardeners are totally devoted to the cause, and they work tirelessly for an often very paltry salary (Dumont 2017).

In addition to these elements, Walloon agriculture does not seem well suited to feeding people. Indeed, most of the production is directed towards exportation; above all, Walloon agricultural production is largely made up of large crops that do not feed the inhabitants. Moreover, both provincial and regional political actors remain attached to intensive and exporting agriculture. Despite the glaring demands of certain segments of the population, this type of agriculture persists and does not seem to be running out of steam. Market gardening in Wallonia is still derisory, and most of the food for the inhabitants of Charleroi and Liège is from supermarkets where agricultural products come from all over the world and so little comes from local farmers. This raises the question of economic, social and environmental benefits and the links between the local and the global. Through the question of fresh, healthy and sustainable food, there is a demand to make citizens aware of the need to reconnect cities materially, socially and economically with the countryside not only for recreation but also for producing food and so preserve and develop both a heritage and an independence from the mainstream global market.

Faced with these civic initiatives, which seek to reclaim their food in the vicinity, bypassing so-called "local" producers, some researchers have underlined the growing interest of supermarket companies that seek to recover this niche claimed by these consumers (Billion 2018). These same researchers have nevertheless observed that in these purchasing groups, consumers opt for food that remains local, on a small scale, passing directly through producers able to meet their expectations. Above all, this approach reconfigures and improves the links between producers and consumers and thus between the city and the surrounding countryside (Morris and Buller 2003). Does this mean that the construction of a network, such as the food land belt, that aims to feed a metropolitan area will necessarily denature the first initiatives and miss some of the aims desired by the initiators?

Many initiatives on food relocation have been identified in the 18 municipalities from the metropolitan areas of Liège and Charleroi. These initiatives were initiated by and involved small segments of the population seeking to feed the population with local production. While urban agriculture is gaining momentum in many cities around the world, the application of the model raises several questions such as the willingness and ability of both the farmers and populations to change the production, marketing and consumption methods because of the affordability of agricultural lands or fresh food prices. The geographical and socio-economic context is also a major constraint that may explain why Liège has more success than Charleroi in the adoption of new food systems. Moreover, researchers converge on the question of whether Walloon agriculture can truly feed the local populations (Dumont 2017; Bousbaine 2020). Finally, a consumer movement focused on proximity and therefore on citizen relations between consumers and producers is developing (Higgins et al. 2008). This means that farmers need to change their orientation to supply the city and find enough land; otherwise, only a segment of the population can benefit from local food. In both cities, we noted the emergence of a new peasantry, people who want to change the world and start farming, who are seen by the sector as dreamers but also slaves according to their poor labour conditions.

If this chapter focuses mainly on the renewed demand of the urban food market, this detour by the city allows us to highlight several challenges that the countryside around cities is facing such as preserving rural landscapes, empowering the local economy, increasing resilience in the face of the world market economy, connecting people and participating in city metabolism as an important factor. After decades of decreasing agricultural activities and the absorption of villages in the metropolitan area as residential districts, there is a chance to recover and strengthen a millennia activity, which may offer a better life for many people and a more sustainable and responsible way of eating.

References

Alexandre, F., and A. Génin. 2014. Ceintures vertes autour des grandes métropoles. In *Formes et échelles des compositions urbaines, Actes des congrès nationaux des sociétés historiques et scientifiques*, ed. P. Pinon, 75–87. Paris: Comité des travaux historiques et scientifiques.

Aubry, C., and J. Pourias. 2013. L'agriculture urbaine fait déjà partie du métabolisme urbain. *Economie et stratégies agricoles, Déméter* 2013: 135–155.

Ba, A., and C. Aubry. 2011. Diversité et durabilité de l'agriculture urbaine: une nécessaire adaptation des concepts. *Norois* 221: 11–24.

Baguette, A. 2015. *La ceinture aliment terre liégeoise: exploration d'un réseau alternatif.* Bruxelles: Entraide et fraternité.

Billion, C. 2018. Rôle des acteurs du commerce et de la distribution dans les processus de gouvernance alimentaire territoriale. Thèse de doctorat en géographie, Université Clermont Auvergne.

Bousbaine, A. D. 2020. Ville et agriculture face à l'émergence des systèmes agro-alimentaires innovants. Etudes de cas dans deux agglomérations wallonnes: Charleroi et Liège. Thèse de doctorat, Université de Liège

Brand, C. 2015. Alimentation et métropolisation; repenser le territoire à l'aulne d'une problématique vitale oubliée. Thèse de doctorat en géographie, Université de Grenoble Alpes.

Chiffoleau, Y. 2008. Les circuits courts de commercialisation en agriculture: diversité et enjeux pour le développement durable. In *Les circuits courts alimentaires. Bien manger dans les territoires*, ed. G. Maréchal, 21–30. Dijon: Educagri éditions.

Dubbeling, M. H. de Zeeuw, and R. van Veenhuizen. 2010. *Cities Poverty and Food. Multi-stake holder Policy and Planning in Urban Agriculture*. Bourton on Dunsmore: Practical Action Publishing.

Dumont, A. 2017. Analyse systémique des conditions de travail et d'emploi dans la production de légumes pour le marché du frais en région wallonne (Belgique), dans une perspective de transition agroécologique. Thèse de doctorat en sciences agronomiques et ingénierie biologique, Université de Louvain-la-Neuve.

Ericksen, P. J. 2008. Conceptualizing Food Systems for Global Environmental Change Research. *Global Environmental Change* 18 (1): 234–245.

FAO. 2003. *Trade Reforms and Food Security: Conceptualizing the Linkages*. Rome: FAO.

Goodman, D. 2004. Rural Europe redux? Reflections on alternative agro-food networks and paradigm change. *Socologia Ruralis* 44 (1): 3.16.

Halleux, J.-M., L. Brück, and N. Mairy. 2002. La périurbanisation résidentielle en Belgique à la lumière des contextes suisse et danois: enracinement, dynamiques centrifuges et régulations collectives. *Belgeo* 2002 (4): 333–354.

Higgins, V., J. Dibden, and C. Cocklin. 2008. Building Alternative Agri-food Networks: Certification, Embeddedness and Agri-environmental Governance. *Journal of Rural Studies* 24 (1): 15–27.

Ilbery, B., M. Kneafsey, and M. Bamford, 2000. Protecting and Promoting Regional Speciality Food and Drink Products in the European Union. *Outlook on Agriculture*, 29 (1), 31–37.

IWEPS (2020). *Chiffres clés de la Wallonie*. Namur: IWEPS.

Janin, P., and S. Dury. 2012. Les nouvelles frontières de la sécurité alimentaire. Une réflexion prospective. *Cahiers Agricultures* 21 (5): 285–292.

Johnson, R., R.A. Aussenberg and C. Tadlock 2013. *The Role of Local Food Systems in U.S. Farm Policy*. Washington: Congressional Research Service.

Louviaux, M. 2011. Le Groupe d'achats communs de Barricade (Liège): à petits pas vers un autre monde. In *La consommation critique, Mouvements pour une alimentation responsable et solidaire*, ed. G. Pleyers, 111–132. Paris: Desclée de Brouwer.

Maréchal, G. 2008. *Les circuits courts alimentaires. Bien manger dans les territoires*. Dijon: Educagri.

Maxwell, S., and R. Slater. 2003. Food Policy Old and New. *Development Policy Review* 21 (5–6): 531–553.

Maye, D., L. Holloway, and M. Kneasfey. 2007. *Alternative Food Geographies. Representation and Practice*. Amsterdam: Elsevier.

Morris, C., and H. Buller, 2003. The local food sector: a preliminary assessment of its form and impact in Gloucestershire. *British Food Journal* 105 (8): 559–566.

Mougeot, L. J. A. 2000. *Urban Agriculture: Definition, Presence, Potentials and Risks*. Ottawa: IDRC.

Moustier, P., and A. Mbaye. 1999. Introduction générale. In *Agriculture périurbaine en Afrique Subsaharienne*, ed. P. Moustier, A. Mbaye, H. De Bon, H. Guérin and J. Pagès, 7–17. Montpellier: CIRAD.

230 Antonia Bousbaine and Serge Schmitz

Mundler, P., J. N. Consalès, G. Melin, C. Pouvesle and P. Vandenbroucke. 2014. Tous agriculteurs? *L'agriculture urbaine et ses frontières*. *Géocarrefour* 89 (1–2): 53–63.

O'Hara, S. U., and S. Stagl. 2001. Global Food Markets and Their Local Alternatives: A Socio-ecological Economic Perspective. *Population and Environment* 22 (6): 533–554.

Poulot, M. 2013. Introduction: Agriculture dans la ville, agriculture pour la ville: acteurs, pratiques et enjeux. *Bulletin de l'association des géographes français* 90 (3): 276–285.

Praly, C., C. Chazoule, C. Delfosse, and P. Mundler. 2014. Les circuits de proximité, cadre d'analyse de la relocalisation des circuits alimentaires. *Géographie, Economie, Société* 16 (4): 455–478.

Prigent-Simonin, A. H., and C. Hérault-Fournier. 2012. *Au plus près de l'assiette. Pérenniser les circuits courts alimentaires*. Versailles: Éditions Quæ.

Rastoin, J. L., and G. Ghersi. 2010. *Le système alimentaire mondial. Concepts et méthodes, analyses et dynamiques*, 584. Versailles: Edition Quae.

Renting, H., T. Marsden, and J. Banks. 2003. Understanding Alternative Food Networks: Exploring the Role of Short Chains in Rural Development. *Environment and Planning* 35 (3): 393–411.

Scheromm, P., C. Perrin, F. Jarrige, L. Laurens, B. Nougaredes, and C. Ruault. 2019. From Ignorance to Commitment: How Periurban Municipalities Deal with Agriculture?. *Geographical Research* 57 (4): 425–435.

Schmitz, S. 2002. Walloon Region: An Agro-forestry Landscape without Farmers?. *Dela* 17: 174–182.

Schmitz, S. 2008. Du new urbanism au new ruralism: un débat en cours sur de nouvelles visions de l'avenir des campagnes aux Etats-Unis. *Géocarrefour* 83 (4): 331–336.

Schmitz, S. 2013. The Prohibition of Dogcarts in Belgium: A Hidden Agricultural Policy?. In *Integration through Subordination. The Politics of Agricultural Modernisation in Industrial Europe*, ed. P. Moser, and T. Varley, 289–299. Turnhout: Brepols.

Theys, J., and E. Vidalenc. 2013. *Repenser les villes dans la société post carbone*. Paris: Ministère de l'Ecologie du Développement durable et de l'Energie, ADEME.

Vachon, B. 1991. *Le Québec Rural dans tous ses états*. Montréal: Boréal,

Van der Ploeg, J. D., and H. Renting. 2004. Behind the redux: a rejoinder to David Goodman. *Sociologia Ruralis* 44 (2): 234–242.

Van Hecke, E., M. Antrop, S. Schmitz, V. Van Eetvelde, and M. Sevenant. 2010. *Paysages, monde rural et Agriculture*, Atlas de Belgique. Gent: Academia Press.

Van Veenhuizen, R. 2006. *Cities Farming for the Future*. Leusden: RUAF Foundation, IIRR and IDRC.

Verhaegen, E. 2011. Le consommateur en tant que consom'acteur. In *la consommation critique. Mouvements pour une alimentation responsable et solidaire*, ed. G. Pleyers, 259–281. Paris: Desclée de Brouwer.

Wiskerke, J. S. C. 2009. On Places Lost and Places Regained: Reflections on the Alternative Food Geography and Sustainable Regional Development. *International Planning Studies* 14 (4): 369–87.

14 Conclusion

Holly Barcus, Roy Jones, and Serge Schmitz

Introduction: rural transformations, adaptations and transitions

Rural transformations have been taking place for millennia, ever since hunter-gatherers began to cultivate crops and domesticate livestock (Barker 2006). However, the pace and scale of these changes began to accelerate as a result of the scientific and transport-related developments that accompanied first the enlightenment and then the Industrial Revolution in the eighteenth and the nineteenth centuries (Jones 2016). While this was an era of imperialism, rather than globalisation (Belich 2009), by the arrival of the twentieth century, it had led to the transformations of rural landscapes on a global scale (Jones and Diniz 2019).

Perhaps inevitably, opinions on these transformations varied. The poet, William Wordsworth, deplored the incursion of the railways, one of the earliest and most powerful forces of globalisation, into the English Lake District (Yoshikawa 2020). Similarly, Short (2006, 177) sees Thomas Hardy's novels of the mid-nineteenth century southern England as:

a picture of the countryside in transition, a rural region rich in local ways and customs, being incorporated into a metropolitan society with national codes and standardised lifestyles. He is, in effect, describing the death throes of an authentically different rural society.

By contrast, Marx and Engels welcomed this transformation and argued, in 'The Manifesto of the Communist Party', that:

(t)he bourgeoisie has subjected the country to the rule of the towns. It has created enormous cities, has greatly increased the urban population as compared with the rural, and has thus rescued a considerable part of the population from the idiocy of rural life.

(cited in Falmer and Jansen 1979, 23)

All of these views have merit since they emphasise the massive cultural and landscape changes that have been and are occurring in the rural world, on the

DOI: 10.4324/9781003110095-17

one hand, and its growing economic productivity (and shrinking demographic significance) on the other.

In the twenty-first century, this complex concatenation of transformations to rural (and urban) landscapes, peoples and economies is accelerating and, according to some observers (e.g. Eriksen 2014), is creating a condition of 'overheating' in more than climatic terms. The wide-ranging economic and social processes of globalisation and urbanisation are significant drivers of these rural transformations (Woods 2005) but, as the case studies in this volume demonstrate, their outcomes are neither single nor simple. These processes have the power to enhance both local diversity and global uniformity and do so in a variety of ways in both the Global North and the Global South. To encompass this complexity, this volume has been structured around three key themes of transformation: agricultural and land use change, or changing places; rural demographic diversity, or changing populations; and rural innovations and rural–urban partnerships, or changing processes. These might also be construed, respectively, as: what changes are taking place in rural areas, who is experiencing these changes, and why these changes are occurring. These themes cannot, therefore, be mutually exclusive and most of the case study chapters document transformations that are relevant to more than one of them.

As a means of integrating our observations on the case studies, we use the lens of Colten's (2019) concept of 'adaptive transitions'. Although he was studying specific issues of coastal management, his approach is nevertheless relevant to the diverse contemporary pressures bringing about the rural transformations described in this volume. Colten identifies both scalar and purposive distinctions between adaptations and transitions. In temporal and spatial terms, he argues that adaptations take place over years and decades and are characteristically local or regional. They are 'human actions taken in response to ... change in an effort to perpetuate a society, even if modified in some form' (Colten 2019, 417). 'Transition', however, is 'a term used by historians to account for multiple adaptions, deliberate or *ad hoc*, coordinated or uncoordinated' (Colten 2019, 417) which result in fundamental change. Such transitions – for example globalisation and urbanisation – take place over centuries and, although they are global, they do not change all places simultaneously or affect all places in the same way. Nevertheless, they 'thoroughly infuse multiple aspects of society and demand social, cultural, political, technological and economic transformations' (Colten 2019, 417).

In this volume, Moseley and Pessereau's study of the increasing use of herbicides by women in a rural region of Burkina Faso provides one illustration of how this single adaptation, which seeks to sustain local agricultural production, links not only to the wider transitions of globalisation and urbanisation but also to our three subthemes of rural transformation. The use of herbicides is a single labour-saving adaptation and, indeed, an innovation. However, this local adaptation is also, at least in part, a response to a wider urbanising and globalising transition whereby, as in many rural regions worldwide, working-age males have left the case study area to seek more profitable urban employment, mainly, in this case, in the country's export-oriented mining industry. This has brought about not only a demographic imbalance in terms of the area's age and sex patterns and, thereby, a rural labour shortage, but also a rural–urban, and even a globalising, partnership through which the chemical products of multinational corporations are now finding their way through ever-expanding global supply chains to this remote part of the international economic periphery. These globally available herbicides and fertilisers have now replaced at least part of the input of human labour which was formerly required to cultivate this land. This single adaptation therefore contributes to agricultural change through the modification of local land use practices and forms a component of Africa's New Green Revolution.

In the following sections of this chapter, we consider the rural transformations documented in the case study chapters under the headings of our three subthemes, both as local and immediate adaptations and as responses and contributions to wider, and even global, transitions.

Agricultural and land use change

All of the chapters on this theme, and most of those in the remainder of the volume, document local changes either in the nature of the agricultural outputs or to the methods by which these outputs are being produced, or both. Robinson et al. describe the shift from grain to horticultural crops in a rural area of Xi'an, China. Kang considers the change from pastoralism to fruit and vegetable production in parts of Turkey and Woods depicts a shift from vegetable (and timber) production to dairying in Tasmania, Australia and from pastoralism to grain (rice and soy) production in Rio Grande do Sul, Brazil. In all these case studies, the shifts in produce types have been accompanied by increases in the use of technology on farms and, in the Turkish case, through the development of regional irrigation systems. Together with the closure of local agricultural processing facilities, such as canning plants in Tasmania and meatworks in Brazil, this has led to reductions in local employment opportunities with labour demands either becoming less overall or, as in the Turkish case, becoming restricted to seasonal peaks.

These case studies therefore present examples, not only of local agricultural change but also of a broader tendency towards what Woods terms 'depeasantisation' as agricultural practices have become more mechanised and commercialised and less labour-intensive and locally oriented. As such, they provide evidence of varied local adaptations to the broader transitions of urbanisation and globalisation. Kamuti's study of Zimbabwe provides a striking contrast to the remainder of the chapters in this section. While there has been no change to the type of agricultural produce (tobacco) being cultivated in this case, the adaptation that has occurred here has been one of what might almost be termed repeasantisation. The local response here has, instead, been to the broader global transition of decolonisation and has taken the form of a rapid land reform process. Here, the agricultural land was formerly organised into large holdings which were owned and managed by white settler colonial farmers. This land has now been reallocated to larger numbers of Zimbabwean smallholders who, given their limited resources, both cultivate and process the tobacco by traditional and more labour-intensive means.

In all these cases, however, a primary focus has been on the land use of agriculture (including livestock farming) and on what Holmes (2006) termed the productivist role of rural land. Nevertheless, Robinson et al. also note the importance of consumption-related rural activities in their depiction of the recent rise in touristic and recreational attractions and developments which has taken place in several of the villages and farms surrounding the city of Xi'an. The role of environmental protection of – at least some – rural land is relevant to the Zimbabwean case study where land use change is occurring in hitherto protected rural woodland as smallholders both increase the area devoted to their tobacco crops and expand their use of native timber as a leaf curing fuel for their tobacco barns.

Agricultural and land use changes also figure in many of the remaining case studies in this volume. Gebrati describes the introduction of new, especially tree, crops in the High Atlas Mountains of Morocco. Barcus and Lanegran describe how new 'tree range' methods of poultry farming have been introduced in peri-urban Minnesota. Several chapters describe the adoption of contrasting, if not oppositional, agricultural methods with regard to the use of agricultural chemicals. Moseley and Pessereau's study of the increase in the use of herbicides in Burkina Faso can be contrasted with Kim et al.'s description of the promotion of more chemical-free farming methods in Vietnam, Tort-Donada and Fumadó-Llambrich's account of an enterprise-based venture in organic farming in Catalonia and Bousbaine and Schmitz's study of similar moves in Belgium. Movements away from productive agriculture towards more consumption-based rural land uses are represented by Gebrati's study of rural tourism in Morocco. Dmochowska-Dudek et al., in their study of built heritage preservation in Polish villages, document how this process has the potential not only to promote the consumption of this rural area by visitors and by urban immigrants, but also to protect a locally distinctive cultural landscape. Barcus and Lanegran also focus on the protection of rural land when they describe how planning regulations in the areas surrounding Minnesota's Twin Cities now seek to preserve local water resources and prevent excessive urban sprawl. Taken together, these case studies showcase the diversity, and even the contradictory nature, of many local adaptations to apparently monolithic transitions such as urbanisation and globalisation.

Rural demographic diversity

As a result of the global transitions discussed above, a wide range of new economic and social opportunities constantly present themselves to populations within and beyond specific rural areas worldwide. The existence of these opportunities encourages members of urban and rural populations, individually or in groups, to move into or out of these regions as they undertake adaptation strategies in order to improve their levels of well-being.

Inward population movements characteristically increase local demographic diversity in ethnic terms. This frequently affects peri-urban areas, as illustrated by Barcus and Lanegran's study of the movement of Hmong and, to a lesser extent, Latinx small farmers into the rural urban interface of Minnesota's Twin Cities. However, it can also take place more widely, as demonstrated by Cawley's study of the near-nationwide movements of Polish and Estonian migrants to many Irish rural towns, where they have taken up a range of employment opportunities in agricultural and food processing occupations. In demographic 'pull' terms, both of these in-movements depict migrants adapting their lives in order to take up local economic opportunities. However, they can also be said to have resulted from 'pushes' provided by global, if not globalising, political transitions. The movement of many Hmong, largely for their own protection, to the United States following the Vietnam conflict was a direct result of the Cold War. More indirectly, it was the ending of that war and the admission of formerly communist Eastern European states to the European Union that has allowed Poles and Estonians to move, relatively freely, to Western European countries such as Ireland.

Inward population movements can also alter the socio-economic composition of rural populations. In rural Poland, Dmochowska-Dudek et al. describe how a demographic shift of more affluent and educated urban dwellers into small and historic rural villages has contributed to both the physical preservation and the social regeneration of these settlements. In Poland, as in Ireland and Minnesota, it was the emigration and resultant ageing of sections of the pre-existing rural populations that, at least in part, provided the physical and economic space for these incoming groups to move into and, thereby, to bring about the environmental, functional and demographic changes which these chapters analyse. The final chapter on this theme, by Moseley and Pessereau, focuses, by contrast, on a side effect of local emigration on the rural population that has been 'left behind'. The movements of younger males from rural regions of Burkina Faso to work in distant mines and cities prompted adaptations in local agricultural practices to make them less physically demanding for the remaining population, which was becoming increasingly female and elderly. But it was the broader transition of globalisation that facilitated both the capital investment in mines in developing nations like Burkina Faso to supply global markets and the development of global reach by multinational corporations, such as those supplying herbicides, pesticides and other agricultural chemicals, to remote rural areas worldwide.

Demographic change and diversification also figure in some of the other case study chapters. Woods' study of agribusiness towns in Brazil and Australia alludes to the outward and (to a lesser extent) inward local population shifts that accompanied the incorporation of these centres into corporate global supply chains. Kang's Turkish case study provides a clear example of demographic diversification, at least on a seasonal basis. She describes how agricultural workers, both from poorer regions elsewhere within Turkey and refugees from neighbouring and war-torn Syria, have adapted to migrant lifestyles as they perform short-term contract work on Anatolian properties that are now being farmed more commercially and intensively as they become increasingly integrated into global production systems. Kang's chapter, like that of Moseley and Pessereau, also focuses on the demographic issue of gender. Both case studies consider how the intersections of local cultural practices with modernising and globalising economic trends can generate severe socio-economic and even health pressures and problems for the women who become caught up in these transformations.

Rural innovations and rural–urban partnerships

Three of the chapters on this theme, those by Kim et al., Bousbaine and Schmitz and Tort-Donada and Fumadó-Llambrich are concerned with a highly specific innovation, namely the supply of 'cleaner' (i.e. less affected by agricultural chemicals) food products to certain sections of the urban market. Although consumers seeking produce of this type, might be considered to be a niche market, the recent growth in demand for such agricultural outputs results from two apparently contrasting global trends, the increasing affluence of many urban dwellers and the growing incidence and level of environmental and health concerns among consumers. The three case studies therefore focus on rural–urban partnerships through which greater quality control and greater attention to environmental safeguards can be provided from the (rural) paddock to the (urban) plate. Two of the chapters focus on relatively local partnerships, on the 'food belts' of Liège and Charleroi in Bousbaine and Schmitz's Belgian study, and on a Catalonian cooperative serving markets in metropolitan Barcelona, as depicted by Tort-Donada and Fumadó-Llambrich. In contrast, Kim et al's. Vietnamese study documents how a national food certification process has facilitated the distribution of certified food products from Lam Dong Province to urban markets throughout the country.

Collectively, these chapters also depict the diversity that exists among the originators of rural innovations. In Catalonia, the rural collective, *Hortec*, developed its own organic production capacity and established niche markets for its outputs in metropolitan Barcelona. In Liège and Charleroi, concerned groups of urban dwellers actively sought out farmers capable of supplying them with environmentally friendly food products from the local 'food belts' surrounding these Belgian cities. In Vietnam, it was the food certification regime, brought in by the national government, that enabled small farmers in Lam Dong to work together to access premium prices for their produce in several urban markets.

Gebrati describes how farmers in the High Atlas Mountains of Morocco have innovated by moving into the cultivation of higher value products, in this case in the form of tree crops, to supply the nation's urban markets in Casablanca and Marrakesh. However, the most notable innovations in this remote rural region have taken place in those villages which have been able to attract various forms of tourist development. This has involved even wider-scale urban–rural partnerships, since both the initiatives for and the tourists participating in this innovation largely originated in Europe.

As has already been implied in the previous sections of this chapter, both rural innovations and rural–urban partnerships have been the key themes in many of the other case studies in this volume. Tourism is both an innovation and the facilitator of rural–urban partnerships in China (Robinson et al.) and Poland (Dmochowska-Dudek et al.) as well as Morocco. The Hmong farmers in Minnesota (Barcus and Lanegran) supply premium produce to local urban markets in a similar manner to those in Catalonia and Belgium. National investment in Turkey (Kang) and international investment in Brazil and Australia (Woods) have resulted in agricultural innovation and in rural–urban partnerships on a global scale.

Conclusion: local and global rural transformations

Colten (2019, 430) acknowledged that, in the management of the Louisiana coastline:

> (m)ost adaptations were reactions to particular situations and not envisioned as a regional plan with a sustainable goal. Some have worked at cross purposes that created fundamental conflicts.

More or less fundamental conflicts of various types have certainly ensued in several of the case studies contained within this volume. Zimbabwean smallholders have contravened environmental regulations in their exploitation of native forests. In Turkey, a transition to a seasonally nomadic form of employment has placed additional strains on traditional, gender-based family structures and fomented disputes between local residents and both Turkish and Syrian migrant workers. In Morocco, a social and economic divide is appearing between those High Atlas villages that are transitioning to tourism and those that are not. Other local adaptations simply seem to be at cross purposes, such as the increase in the use of agricultural chemicals in Burkina Faso and their decrease in parts of Vietnam or the simultaneous modernisation of farming and heritagisation of rural tourism in both China and Poland.

However, these local transformations hardly provide support for Marx and Engels' contention of the idiocy of rural life. What this collection of case studies shows is that, while rural dwellers are, increasingly, required to adapt locally to external trends and pressures, how they undertake these adaptations may be neither simple nor obvious. As Soares and Collins (1982, 57) argue – in a statement that is not only applicable to peasants:

> Peasants are not prisoners of their personality, of their culture, nor of a specific mode of production; they do not react in the same way everywhere and at all times. Their reaction has been varied throughout time and space.

In supporting this view, Eriksen (2014, viii) echoes Colten when he contends that 'what we are confronted with is a series of *clashing scales* (emphasis in the original) which remain poorly understood'. Of particular relevance to a major theme of this volume, he states:

> Many widely read authors writing about the interconnected world seem to be hovering above the planet in a helicopter with a pair of binoculars. They may get the general picture right but fail to see the nooks and crannies where people live (Eriksen 2014, viii).

Through this series of case studies, we have shone a light into the diverse 'nooks and crannies' of several corners of the rural world. In doing so, we have sought to show how the people who live there (and many people who do not) all contribute to what is a highly complex 'general picture' of contemporary rural transformation.

References

Barker, G. 2006. *The Agricultural Revolution in Prehistory: Why Did Foragers Become Farmers?* Oxford: Oxford University Press.

Belich, J. 2009. *Replenishing the Earth: The Settler Revolution and Rise of the Anglo-World 1783-1989.* Oxford: Oxford University Press.

Colten, C. 2019. Adaptive Transitions: The Long-term Perspective on Humans in Changing Coastal Settings. *Geographical Review* 109 (3): 416–435.

Eriksen, T. 2014. *Overheating: An Anthropology of Accelerated Change.* London: Pluto Press.

Falmer, H. and J. Jansen, eds. 1979. *Spatial Inequalities and Regional Development.* Dordrecht: Springer.

Holmes, J. 2006. Impulses towards a Multifunctional Rural Transition: Gaps in the Research Agenda. *Journal of Rural Studies* 22: 142–160.

Jones, P. 2016. *Agricultural Enlightenment: Knowledge, Technology, and Nature, 1750–1840.* New York: Oxford University Press.

Jones, R. and A. Diniz eds. 2019. *Twentieth Century Land Settlement Schemes.* London and New York: Routledge.

Short, J. 2006. *Imagined Country: Environment, Culture and Society.* New York: Syracuse University Press.

Soares, G. and J. Collins. 1982. The Idiocy of Rural Life. *Civilisations* 32 (1): 31–65.

Woods, M. 2005. *Rural Geography.* London: Sage.

Yoshikawa, S. 2020. *William Wordsworth and Modern Travel: Railways, Motorcars and the Lake District, 1830-1940.* Liverpool: Liverpool University Press.

Index

Page numbers in **bold** indicate tables, page numbers in *italic* indicate figures.

adaptive transitions 232
African farming systems, women's labor in 153
Aggregate Rural Areas (ARAs) 119, 124
Aggregate Town Areas (ATAs) 119, 124
agrarian culture in Catalonia 171–173
agribusiness 16; as a collective noun 17; corporations, in Smithton 4, 22, 25; financing from companies of 19, 20; fiscal contribution of 19; hegemony 27, 29; investment 19, 23; local farmers and transnational agribusiness relationship in 21; pro-agribusiness regimes 28; profit maximization by 17; service providers 21, 22, 26; as a singular noun 17; transnational 15, 18, 21, 28, 29
agribusiness boom 26
agribusiness city 16, 18, 29
agribusiness-friendly projects 25
agribusiness fringes 21, 25, 26
agribusiness towns 4, 16, 18, 29; Dom Pedrito, in Rio Grande do Sul 19–22; Smithton, in Tasmania 22–26
agricultural and land use change 4, 232–234
agricultural hyper productivity 2
Agricultural Preserve 101
agricultural production 2, 18, 29, 32–34, 75, 81, 98, 138, 153, 175, 193, 227, 232
agricultural products 32, 62, 63, 101, 176, 200, 204
agricultural transition in Rural China 53; "Farmhouse Joy" tourism 63–65; multifunctional agriculture, transition to 57–63; policy setting 56, 57; theorising 54, 55
agricultural transitions 3, 16
agricultural value 102, 202

agricultural workers: migration, in Şanlıurfa Province 39, *39*; seasonal 32–39, 41, 44, 46, **47**, 49, 50
agricultural working conditions of female workers 46–48
Agriculture 4.0 55
agriculture; *see also* agribusiness; dairy farming; corporate 16; digital 17; financialization of 17; globalization of 15, 17; productivist 3
agri-food production 15, 17
agrochemical industry 152
agroforestry 81
American Community Survey 95
Aquaculture Stewardship Council (ASC) Standards 201
arable farming growth, in Dom Pedrito 19
artisanal gold mining, rural labor dynamics and 157, 158
Asni Township Nut Festival 193, 194
Association of Religious Data Archives (The ARDA 2020) 95
atrazine 156, *156*
Azzaden Valley 189

Bai Luyuan 64
Bailuyuan – Bailucang scenic area 64, *64*
Barcus, H. R. 94, 113, 118, 128
BASF 20
Bayer 20, 156
Berger, J. 171
Bernard, M. 188
Berriane, M. 188
Best Aquaculture Practices (BAP) 201
birth rate, in Şanlıurfa Province 43, 44, *44*
Biskupian folk culture 132, *140*
Biskupizna folk culture 132, 137, 138, 142
Bobo Dioulasso 154–156

Boujrouf, S. 193
bovine spongiform encephalopathy
   (BSE) 179
Brachystegia 73
bread 41
bricolage 74–76, 81, 82
British American Tobacco 79
Brouder, P. 9
Brugger, F. 158
Buddhist temple 99
bug hotel 142
Burkina Faso 6, 7, 151, 152, 157, 160,
   232, 234, 235, 237; Government of
   Burkina Faso (2016) 158; supply chain
   to 154, 155–157

Carney, J. 153
Catalonia, Spain: agrarian culture in
   171–173; innovative, versatile and
   permanently adaptive figure of *pagès*
   174; *pagès*/Catalan farmer, role played
   by 174; *pagès* as the "cornerstone" of
   Catalan society 175
Ceinture Aliment Terre 222
Charleroi 219, 221, 223, 228
Charleroi Métropole 223, 224
Charoen Pokphand Group 201
cherry production in rural China 60, **61**
China, agricultural transition in 53;
   "Farmhouse Joy" tourism 63–65;
   modernising agriculture 59–63;
   multifunctional agriculture, transition
   to 57–63; policy setting 56, 57;
   theorising 54, 55
China National Development and Reform
   Commission 64
Chinese Communist Party 56
Choropleth mapping 113, 119, 128
CILSS regional system 163
Circular Head 22, 23
Clapp, J. 152
Cleaver, F. 74, 75
closed traditional society, female workers
   in 43–45
cluster analysis 118, 119, 120, 121, 126,
   126, 128
Cold War 235
Collins, J. 237
Colophospermum mopane 73
Colten, C. 232, 237
*Comité Permanent Inter-Etats de Lutte
   contre la Sécheresse dans le Sahel* (CILSS)
   156, 157
Communal Land Forest Produce Act
   76, 82

communal land tenure in Zimbabwe 72, 74
communications technologies 5
community-led local development
   (CLLD) 146
Community-Supported Agriculture
   (CSAs) 95
Constitution of Zimbabwe, 76, 81, 82
consumption-focused developments 18
contract-based vegetable cultivation
   206–209
contract farmers 78
conventional farmers becoming VietGAP
   producers 209, 210
Cooperativa Hortec 10, 170, 175–180; for
   future generations of Catalan farmers
   180, 182; historical significance of 181,
   182; "innovative" agro-productive
   model of 180–182
cooperativism 176, 177, 179
Copán Valley of Honduras 80
corporate agriculture 16
corporate concentration 17, 152
corporatization of agricultural systems 17
Council of Europe (CoE) 45
County Board 100
County Tax Assessor 101
COVID-19 pandemic 3, 221
crop management 21
CSO (Central Statistics Office) 115, 121
Çukurkuyu town in Niğde Province 36, 37

dairy farming; *see also* agriculture; in
   Circular Head 23; conversion of land
   to 26; in New Zealand 23; in Smithton
   23, 28
Dakota County, Rural–Urban Interface
   (RUI) in 91, 92, 95–98; changing
   ethnoracial and cultural diversity in
   RUI 108; conceptual framework 92;
   connection of local to global scale 107,
   108; contested RUI 92, 93; Dakota
   County, Minnesota (case study)
   95–98; demographic change in 98–100;
   farmland preservation 93, 94; farmland
   protection and comprehensive land
   protection plan 100–102; HAFA farms
   104–106; Latinx population 96, 97;
   Mainstreet Project 106; methods and
   approach 94, 95; new farm operators
   and agricultural entrepreneurs 102–104;
   non-White population 97; percent
   change in farm operators (2002–2017)
   103; population growth (1900–2017)
   96; social-economic and environmental
   impacts 107

Dakota County Farmland and Natural Area Protection Plan 98
deforestation 71, 73, 75–77, 83; and environmental concerns due to tobacco production 79–81
demographic change 5, 7, 92, 94, 96, 102, 104, 109; in Dakota County 98–100; and diversification 235
depeasantization 5, 17
digital agriculture 17
disintermediation 202
Domachowo: *Biskupian* folk culture in the public space of *140*; centres of the village of *137*; rural public spaces and their perception 135–138; territorial context and characteristics of 132, 133
Dom Pedrito, in Rio Grande do Sul 19–22; arable farming growth in 19; farmland use in 20; food processing sector in 21, 22; inequalities between farmers in 21, 22; local economy 19; location of 19; pro-agribusiness regimes of 28; reinvention, as an agribusiness town 22; rice farming in 19; soil conditions around 20; soybean cultivation in 19, 21, *27*; transnational agribusiness in 21, 28; unemployment in 26
Dowd-Uribe, B. 152
dragonheads 53, 56
Dublin 118, 119, 121
Duck River Co-op Butter Factory in Smithton 22, 23
Duc Trong district, VietGAP certified area in 203, 204, *205*
Dutch Mill 24

e-commerce 62, 63, 67, 224
economic independence of female workers 34, 35, 39, 46–48, *48*
Economic Interest Grouping (EIG) 194, *194*
educational levels of female workers 38, *44*
education level of female workers in Turkey 43
Electoral Districts (EDs) 118, 119, 121, 122, 125, 128, 129
electronic cottagers 3
Elias, D. 16, 18, 21, 26, 29
environmental conservation areas 4
Environmental Management Act 76, 82, 83
Eriksen, T. 237
European Union (EU) 8, 55, 113, 143
export-oriented commodity agricultural production 29

family-run processing companies 15
farmers' cooperatives 15, 56
farmers' foods 63
farm-gate sales of cherries 62
farm households 55
"Farmhouse Joy" tourism 63–65
Farmland and Natural Areas Program (FNAP) plan 98, 100, 107
farmland in peri-urban fringe 57
farmland preservation in Dakota County 93, 94, 102
farmland protection and comprehensive land protection plan 100–102
farmland use in Dom Pedrito 20
farm-to-table movement 8, 94
fast-track land redistribution program 73, 74
fast-track land reform program 71–73, 77, 82, 83
"fast track" programme 5
female workers in Southeastern Anatolia 32; agricultural working conditions of female workers 46–48; in closed traditional society 43–45; economic independence of female workers 39, 46–48, *48*; geographical mobility routes for livelihood of seasonal workers 39–43; methodological considerations 36–39; seasonal agricultural activities 40; seasonal agricultural workers 32, 34, 35; tea culture, patriarchal culture examined through 38, 46
female workers in Şanlıurfa Province: daily wages of *47*; educational levels of *44*
feminist political ecology of West Africa's herbicide revolution 151; Burkina Faso, supply chain to 154, 155–157; global economic changes and agrochemical industry 152; global herbicide industry, changes in 155–157; household level labor and gender dynamics 158–162; methods 154, 155; New Green Revolution for Africa (GR4A) 153; policy recommendations 162, 163; rural labor dynamics and artisanal gold mining 157, 158; women's labor in African farming systems 153
Ferme Larock 221
fibre-optic networks 25
financing from agribusiness companies 20
firewood 79, 80, 81
flue-cured tobacco sales 77, **78**
Food Land Belt of Charleroi Métropole 224

food land belts 219, 222–224
food production and consumption 9
food purchasing groups in Charleroi and Liège 220, 221
food safety 10, 200, 201
Forest (Control of Firewood, Timber and Forest Produce) Regulations of 2012 76
Forest Act 76, 82
Forestry Commission 81, 83
French Alpine Club 191
French Protectorate of Morocco in 1912 192
fresh food, reshaped demand for 220; food land belts 222–224; food purchasing groups in Charleroi and Liège 220, 221; health crisis as triggering factor 220
Fumadó-Llambrich, Jordi 10
Furuseth, O. J. 93

Gamisans, Josep Maria 177
"gastro tourism" attractions 189
gaucho culture of the ranchers 19
gender dynamics, household level labor and 158–162
General Directorate of State of Hydraulic Works 34
gentrification 6
geographical indication (GI) 194
geographical mobility routes for livelihood of seasonal workers 39–43
geographic intelligence 171
global farmlands 4, 15, 17, 18, 29
GLOBALGAP 201
global herbicide industry, changes in 151, 154, 155–157
Global Middle 2
Global North 2, 9, 18, 152, 157, 217, 232
global rural transformations 1, 237
Global South 2, 10, 153, 157, 162, 232
glyphosate 7, 156, *156*, 157, 162
Goat Market 135, 137, 138
golden leaf 71
gold mining 157, 158
Great Plains of USA 113
Green Acres program 101
greenbelts 93
Green Revolution 153
Green villages of Biskupizna 141
Guidelines for the Construction of the Beautiful Village 65

HAFA (Hmong American Farmers Association) 104–106
Haggblade, S. 152, 157, 159
Halfacree, K. 135

hand weeding 158, 159, 160
Hardy, Thomas 231
hauliers 22
health crisis as triggering factor 220
hegemony of agribusiness 27, 29
herbicides application 158, 159; backpack sprayers 160; having men apply for 159, 160, 161; impacting children through nursing 161; in polygamous households 161; time savings associated with using herbicides 159; by women 161, **161**
Hiawatha Valley Farm Cooperative (HVFC) 105
High Atlas Mountains of Morocco 234, 236
High Atlas of Marrakesh 10, 186, 189–191; mountain tourism 191–193; relationships between rural and urban areas 195, 196; rural tourism 188, 189; territorial resources, activation of 193–195
Hmong American Farmers Association (HAFA) 104
Hmong farms 105
Hmong refugees in 1976 105
Hoi Toan Cooperative 212
"Hollowed" villages 53
Holmes, J. 2, 3, 234
Hoover Index (HI) 118–120, 124, **124**, 129
household chores: men's likelihood of sharing 38; women engaged in 36
household level labor and gender dynamics 158–162
Household Responsibility Scheme (1980) 4
household responsibility system (HRS) 56, 59, 66
Hurungwe District 81

immigrant labour, in Europe 113, 114
imperialism 2, 231
independent municipality 18
index of dissimilarity (ID) 113, 120, 125, **125**, 128, 129
Indigenous groups 3
Indigenous lands 4
Indigenous protected areas 4
Innovative Alliance for Sustainable Development 222
institutional bricolage 75, 76, 81, 82, 84
institutional dynamics related to forest resources and tobacco production 81–84
International Labor Organization (ILO) 45

investment, agribusiness 19, 23, 26
inward population movements 234, 235
Inwood, S. M. 92
Ireland, labour immigration in 114–118,
    115, *117*; counties in Ireland *117*; data
    sources and methods of analysis 118–
    121; distribution of clusters of EDs *127*;
    distribution of Polish nationals (2016)
    123; Lithuanian and Polish nationals
    distribution (2006–2016) 121–128
Isoberlinia species 73

Johnson, K. M. 113, 120
Julbernardia 73
July 2019 27th Annual Colloquium of
    the International Geographical Union
    Commission on the Sustainability of
    Rural Systems 1

Kautsky, K. 54, 67
Kurangwa, W. 79

Labor Law of Turkey 45
labour immigration and demographic
    transformation 113; data sources
    and methods of analysis 118–121;
    Lithuanian and Polish nationals
    distribution (2006–2016) 121–128;
    migrant labour immigration in Ireland
    114–118, **115**, *117*
Lam Dong Province, VietGAP program
    in, *see* VietGAP (Vietnamese Good
    Agricultural Practices)
Land Conservation Agricultural Land
    Stewardship Program 100
Land Conservation Natural Area
    Protection Program 100
landowners 19, 22
land ownership 32, 33, 34, 147
land use dynamism 5
land use transitions 2, 4
Lapping, M. 93
Latinx population 96, 97
LEADER programme 8, 219
Lefebvre, H. 135
Lenin, V. I. 54
Les petits producteurs 221, 224
Lichter, D. T. 113, 120
Liège, dairy farming in 219
Liège Food Land Belt 222, 223
Lithuanian and Polish nationals
    distribution (2006–2016) 121–123;
    measuring 124, 125; relationships
    between 126–128
livestock ranching 19, 21

local rural transformations 237
Loess Plateau 219
Lowder, S. K. 59
Luna, J. K. 152
Luís Eduardo Magalhães 18

Main Street Project 104, 106
Mak, G. 171
Manyanhaire, I. O. 79
Mapedza, E. 82
Marrakesh, High Atlas of 186,
        189–191; mountain tourism 191–193;
        relationships between rural and urban
        areas 195–196; rural tourism 188,
        189; territorial resources, activation of
        193–195
Marrakesh Tourist Office 191
Marsden, T. 7
Massey, D. 135
McCain 25, 26
Mediterranean countries, sustainable
        agriculture in 169; Catalonia as a
        paradigm of the problems 171–175;
        Cooperativa Hortec, case study of
        175–182; theoretical insights 170, 171
METIP project 45
Metropolitan Agricultural Preserve
        Program 101
Metropolitan Council 101
middlemen system 202
Middle West 6
migrant female workers 114
migrant labour force 5
migrant labour immigration in Ireland
        114–118, **115**, *117*
migrant workers 39, 114
migration of agricultural workers in
        Şanlıurfa Province 39, *39*
milk supply, in Smithton 22
Ministry of Agriculture and Rural
        Affairs of People's Republic of China
        (MARA) 64
Ministry of Agriculture and the High
        Atlas Foundation 194
Ministry of Environment, Water and
        Climate 82, 83
Ministry of Lands, Water and
        Agriculture 83
Minnesota Agricultural Enterprise for
        New Americans (MAENA) 105
Minnesota Grown 96
Minnesota Legislature in 1980 101
Miombo woodlands 73, 79
Mitsubishi 25
Moon Lake Investments 24, *24*

Moroccan–French cooperative venture 191
Morocco 188, 191, 192
mountain tourism 188, 191–193
multifunctional agriculture (MFA) 57; modernising agriculture 59–63
multifunctionality 55, 67, 145, 171
Murray Goulburn 25, 26
Musoni, S. 79

Napoleonic Code in Europe 56
National Tourism Administration 63
neo-endogenous structures 6, 8
neo-liberal national policies in the 1980s 32
neo-productivism 55
new farm operators and agricultural entrepreneurs 102–104
New Green Revolution 7
New Green Revolution for Africa (GR4A) 153
New Zealand, dairy farming in 23
niche farming 98, 105, 107
Niğde Province, Çukurkuyu town in 36

One Belt One Road initiative, in China 64
One-Child Policy 59
Oued Zat Basin 194, 195
Ourika Valley 189, 191

pages/Catalan farmer: as the "cornerstone" of Catalan society 175; innovative, versatile and permanently adaptive figure of 174; role played by 174
paid employment 32
paraphernalia 27
participative store 221
patriarchal culture 38, 48, 49; examined through tea culture 46; women in patriarchal society 35
Pearson correlation 118, 121, 126
Pequeno, R. 16, 18, 21, 26, 29
periphery, rural 91; changing ethnoracial and cultural diversity in RUI 108; conceptual framework 92; connection of local to global scale 107, 108; Dakota County, Minnesota (case study) 95–98; demographic change in Dakota County 98–100; farmland preservation 93, 94; farmland protection and comprehensive land protection plan 100–102; HAFA farms 104–106; Mainstreet Project 106; methods and approach 94, 95; new farm operators and agricultural entrepreneurs

102–104; Rural–Urban Interface (RUI), contested 92, 93; social-economic and environmental impacts 107
peri-urban fringe, farmland in 55, 57
permanently agriculture 101
Pham, V. H. 202
Philip Morris 79
pick-your-own (PYO) schemes 57
Pla, J. 170, 171, 173–175
pluri-active livelihoods 19
Point Ferme 221
Pokorny, B. C. 158
Poland's Voivodeship (province-region) of Wielkopolskie 134
policymaking 76
polygamous households, herbicide application in 161, 162
polygamy, in Turkey 35, 43
post-Fordist tourism 188
post-productivism 2, 55, 171
Potarzyca: rural public spaces and their perception 139–141; spatial development pattern in *140*; territorial context and characteristics of 132, 133
primary education in Turkey 43
pro-agribusiness regimes 28
productivism 2, 4
productivist 3, 54
profit maximization by agribusinesses 17
The Project for the Improvement of the Working and Living Conditions Lives of Seasonal Migratory Agricultural Workers 35
public spaces, shaping 132; anatomy of the action 142; community initiatives 141, 142; intermediate outcomes and causal pathways 143; internal assumptions 143–145; rural public spaces and their perception 135–141; sources of information and methodology of research 134, 135; territorial context and characteristics of three villages 132, 133; theoretical context 133, 134
Purchase of Development Rights (PDR) 93

Religious Congregations and Membership Study (2010) 95
Rhérhaya Valley 189, 191, 192, 193
rice farming, in Dom Pedrito 19
Robbins, P. 94
Robbins, R. 152
Roep, D. 7
Roundup herbicide 28
rural demographic diversity 234, 235

rural innovations 236
rural resources, revaluation of 2
rural tourism 4, 6, 8, 53, 63, 186, 188–192, 234, 237
Rural–Urban Interface (RUI), *see* Dakota County, Rural–Urban Interface (RUI) in
rural–urban partnerships 236
rural web 7
Rye, J. F. 113

safe vegetables 200, 201
safe vegetables production by VietGAP buyers 211; horizontal integration 212, 213; vertical integration, case of 212
Saint Paul Farmers Market 98
Saint Paul-Minneapolis Minnesota metropolitan area 95
Sanders, S. R. 183
sanitary products, purchase of 39, 48
şanlıurfa Province 33, 36–38, 43, 44
Santa Maria River 28
Saputo 25
school-age children of seasonal workers 45
seasonal agricultural activities in şanlıurfa Province 40
seasonal agricultural workers: living conditions of the families of 42; in Southeastern Turkey 32, 34, 35; temporary living conditions of 42
seasonal labour demands 114
seasonal workers 34; geographical mobility routes for livelihood of 39–43; school-age children of 45
settler colonialism 16
Sharp, J. S. 92
sheep-farming operations 24
Short, J. 231
short food supply chains 171
Shroeder, R. A. 153
Silk Road 64
Simmons, L. 94, 113, 118, 128
Simpson Index 102, *104*
Slavíková, L. 84
Slettebak, M. H. 113
smallholders 5, 21
smallholder tobacco production in Zimbabwe 71, 77–79
"small or pocket-sized" square 138
small-scale family farming 17
small-scale tobacco production 76
Smithton, in Tasmania 22–26; agribusiness corporations in 25; agribusiness hegemony in 27; dairy farming in 23, 28; Duck River

Co-op Butter Factory in 22, 23; food processing in 25; location 22; milk supply in 22; pro-agribusiness regimes of 28
Soares, G. 237
social innovation 8
socially vulnerable rural women 34
socio-spatial inequalities 18
"soft" EU programmes 143
Soja, E. W. 135
Soria, A. 172
Southeastern Anatolia, Turkey, women in 32; agricultural working conditions of female workers 46–48; economic independence of female workers 39, 46–48, *48*; female seasonal agricultural workers 32, 34, 35; female workers in closed traditional society 43–45; geographical mobility routes for livelihood of seasonal workers 39–43; methodological considerations 36–39; seasonal agricultural activities **40**; tea culture, patriarchal culture examined through 38, 46
Southwestern Burkina Faso 154, 155, *155*, 162
soybean cultivation, in Dom Pedrito 19–21, 27
soy monoculture 26
spatiality 135
Stara Krobia: rural public spaces and their perception 138, 139; spatial development pattern in *139*; territorial context and characteristics of 132, *133*
state-sponsored, agribusiness-friendly projects 25
Statistical Package for the Social Sciences 120
supermarketization 201
super-productivism 55
sustainable agriculture in Mediterranean countries 169; Catalonia as a paradigm of the problems 171–175; *Cooperativa Hortec*, case study of 175–182; theoretical insights 170, 171
Swiss Civil Code in 1926 35

tax calculation, in Dakota County 102
Taylor, L. 92
tea culture, in Turkey 38, 46
Terluin, I. J. 146
territorial innovation 195, 196
territorialism 9
territorial resources, activation of 193–195
Theory of Change (ToC) approach 135

Thrift, N. 135
Tien Huy Cooperative 212
Tobacco Industry Marketing Board 78, 79
tobacco production in Zimbabwe 71;
    deforestation and environmental
    concerns due to 79–81; forest
    institutions, theoretical framework
    of 74–76; institutional dynamics
    related to forest resources and 81–84;
    methodological approach 76, 77;
    smallholder tobacco production in
    Zimbabwe 77–79
Toilet Revolution 65
Torres, Salomó 177
total fertility rate (TFR) in şanlıurfa
    Province 43, 44
Toubkal Massif 191
Toubkal National Park 191, 192, 194
tourism 196; defined 188; rural 188, 189
tourist village 64
traditional farming towns 15, 29
Tran, N. 202
Transfer of Development Rights (TDR) 93
transnational agribusiness 4, 15, 18, 21,
    28, 29
transnational business 16
transnational networks 15
transnational seed producer 20
tribal groups 33
Turkish Statistical Institute 43
Twin Cities Metropolitan Area 101
Twin Cities residents 98

unemployment: in Dom Pedrito 26; in
    Ireland 118
United Nations World Tourism
    Organization 188
unpaid family workers 48
urban agriculture 226, 228
urban growth buffers 93
urbanization 18, 21, 100, 121, 201
urban–suburban–rural identities 106
U. S. Department of Agriculture 102

Van der Ploeg, J. D. 7
Van Diemen's Land Company (VDLC)
    22, 24, 24, 27
Vang, C. Y. 105
vegetable cannery in Australia 23
vegetable cultivation 204, 205; contract-
    based 206–209; technical advice for 208
vegetable dehydration factory in
    Australia 23
Vicens-Vives, J. 173, 175
VietGAP (Vietnamese Good Agricultural
    Practices) 200, 205; access to 209, 209;

characteristics of VietGAP farmers 205,
    206; contract and price guarantee 207;
    contract-based vegetable cultivation
    206–209; conventional farmers
    switch to become VietGAP producers
    209, 210; crop choice decision and
    harvesting process 207; revenue
    change 208; safe vegetables production
    by VietGAP buyers 211–213; study
    area 204–209; vegetable cultivation,
    technical advice for 208
Vietnam War 105
Vilar, P. 172, 175
Village Housewives Circle 137

wages of women workers in Turkey 47
Wallonia 218, 219
Walloon agriculture, in Belgium 217, 227,
    228; fresh food, reshaped demand for
    220–224; perception and the role of the
    farmers 224–226
Walmart supermarket 21
Ward's method 121
weeding 159, 160
West Africa's herbicide revolution,
    feminist political ecology of 151;
    Burkina Faso, supply chain to 154,
    155–157; global economic changes
    and agrochemical industry 152; global
    herbicide industry, changes in 155–157;
    household level labor and gender
    dynamics 158–162; methods 154–155;
    New Green Revolution for Africa
    (GR4A) 153; policy recommendations
    162–163; rural labor dynamics and
    artisanal gold mining 157–158; women's
    labor in African farming systems 153
Western Development Strategy (WDS)
    programme 64
White Deer Plain 57–58, 58, 58, 59, 60,
    61, 63, 65, 66
Whitlow, J. R. 84
Wilson, G. A. 55
Wiskerke, J. S. C. 7
Wójcik, M. 132
women: desire for economic independence
    48; engaged in household chores
    36; herbicides application in fields
    158–161; labor in African farming
    systems 153; paid labor for hand
    weeding 159; in patriarchal society 35;
    socially vulnerable rural women 34;
    in Southeastern Anatolia, *see* female
    workers in Southeastern Anatolia;
    wages of women workers in Turkey 47
wood consumption 79

wood-fueled tobacco barn 79
Woods, M. 1, 4, 7, 113, 132, 233, 235
Woolnorth 22, 24–25
Wordsworth, William 231
World Heritage sites 4

Xi'an municipal government 65

Yin, R. 95
Yufka 39

Zanetti, J. 158
Zat Valley 189
Zimbabwe, tobacco production in 71;
    deforestation and environmental
    concerns due to 79–81; forest
    institutions, theoretical framework
    of 74–76; institutional dynamics
    related to forest resources and 81–84;
    methodological approach 76, 77;
    smallholder tobacco production 77–79

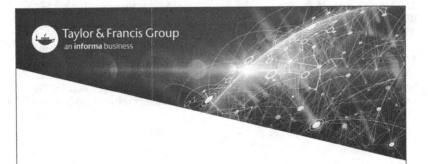

Printed in the United States
by Baker & Taylor Publisher Services

Printed in the United States
by Baker & Taylor Publisher Services